**BROOKLANDS BOOKS**

# LAND ROVER
## SERIES II & IIa
## 1958-1971

Compiled by
R.M. Clarke

ISBN 0 948207 98 1

Booklands Books Ltd.
PO Box 146, Cobham, KT11 1LG
Surrey, England

Printed in Hong Kong

# BROOKLANDS BOOKS

## BROOKLANDS ROAD TEST SERIES
AC Ace & Aceca 1953-1983
Alfa Romeo Alfasud 1972-1984
Alfa Romeo Alfetta Coupes GT. GTV. GTV6 1974-1987
Alfa Romeo Giulia Berlinas 1962-1976
Alfa Romeo Giulia Coupes Gold Portfolio 1963-1976
Alfa Romeo Giulia Coupes 1963-1976
Alfa Romeo Giulietta Gold Portfolio 1954-1965
Alfa Romeo Spider Gold Portfolio 1966-1991
Alfa Romeo Spider 1966-1990
Allard Gold Portfolio 1937-1959
Alvis Gold Portfolio 1919-1967
American Motors Muscle Cars 1966-1970
Armstrong Siddeley Gold Portfolio 1945-1960
Aston Martin Gold Portfolio 1972-1985
Austin Seven 1922-1982
Austin A30 & A35 1951-1962
Austin Healey 100 & 100/6 Gold Portfolio 1952-1959
Austin Healey 3000 Gold Portfolio 1959-1967
Austin Healey Sprite 1958-1971
Avanti 1962-1990
BMW Six Cylinder Coupes 1969-1975
BMW 1600 Col. 1 1966-1981
BMW 2002 1968-1976
BMW 316, 318, 320 Gold Portfolio 1975-1990
BMW 320, 323, 325 Gold Portfolio 1977-1990
Buick Automobiles 1947-1960
Buick Muscle Cars 1965-1970
Buick Riviera 1963-1978
Cadillac Automobiles 1949-1959
Cadillac Automobiles 1960-1969
Cadillac Eldorado 1967-1978
High Performance Capris Gold Portfolio 1969-1987
Chevrolet Camaro SS & Z28 1966-1973
Chevrolet Camaro & Z-28 1973-1981
High Performance Camaros 1982-1988
Camaro Muscle Portfolio 1967-1973
Chevrolet 1955-1957
Chevrolet Corvair 1959-1969
Chevrolet Impala & SS 1958-1971
Chevrolet Muscle Cars 1966-1971
Chevelle and SS 1964-1972
Chevy Blazer 1969-1981
Chevy EL Camino & SS 1959-1987
Chevy II Nova & SS 1962-1973
Chrysler 300 Gold Portfolio 1955-1970
Citroen Traction Avant Gold Portfolio 1934-1957
Çitroen DS & ID 1955-1975
Citroen SM 1970-1975
Citroen 2CV 1949-1988
Shelby Cobra Gold Portfolio 1962-1969
Cobras and Cobra Replicas Gold Portfolio 1962-1989
Cobras & Replicas 1962-1983
Chevrolet Corvette Gold Portfolio 1953 1962
Corvette Stingray Gold Portfolio 1963-1967
Chevrolet Corvettee Gold Portfolio 1968-1977
High Performance Corvettes 1983-1989
Daimler SP250 Sport & V-8250 Saloon Gold Portfolio 1959-1969
Datsun 240Z 1970-1973
Datsun 280Z & ZX 1975-1983
De Tomaso Collection No.1 1962-1981
Dodge Charger 1966-1974
Dodge Muscle Cars 1967-1970
Excalibur Collection No.1 1952-1981
Facel Vega 1954-1964
Ferrari Cars 1946-1956
Ferrari Dino 1965-1974
Ferrari Dino 308 1974-1979
Ferrari 308 & Mondial 1980-1984
Ferrari Collection No.1 1960-1970
Fiat-Bertone X1/9 1973-1988
Fiat Pininfarina 124 + 2000 Spider 1968-1985
Ford Automobiles 1949-1959
Ford Bronco 1966-1977
Ford Bronco 1978-1988
Ford Consul. Zephyr Zodiac MkI & II 1950-1962
Ford Cortina 1600E & GT 1967-1970
Ford Fairlane 1955-1970
Ford Falcon 1960-1970
Ford GT40 Gold Portfolio 1964-1987
Ford RS Escorts 1968-1980
Ford Zephyr Zodiac Executive MkIII & MkIV 1962-1971
High Performance Capris Gold Portfolio 1969-1987
High Performance Escorts Mk1 1968-1974
High Performance Escorts Mk II 1975-1980
High Performance Escorts 1980-1985
High Performance Escorts 1985-1990
High Performance Fiestas 1979-1991
High Performance Mustangs 1982-1988
Holden 1948-1962
Honda CRX 1983-1987
Hudson & Railton 1936-1940
Jaguar and SS Gold Portfolio 1931-1951
Jaguar XK120 XK140 XK150 Gold Portfolio 1948-1960
Jaguar MkVII VIII IX X 420 Gold Portfolio 1950-1970
Jaguar Cars 1961-1964
Jaguar Mk2 1959-1969
Jaguar E-Type Gold Portfolio 1961-1971
Jaguar E-Type 1966-1971
Jaguar E-Type V-12 1971-1975
Jaguar XJ12 XJ5.3 V12 Glold Portfolio 1972-1990
Jaguar XJ6 Series II 1973-1979
Jaguar XJ6 Series III 1979-1986
Jaguar XJS Gold Portfolio 1975-1990
Jeep CJ5 & CJ6 1960-1976
Jeep CJ5 & CJ7 1976-1986
Jensen Cars 1946-1967
Jensen Cars 1967-1979
Jensen Interceptor Gold Portfolio 1966-1986
Jensen Healey 1972-1976
Lamborghini Cars 1964-1970
Lamborghini Countach Col No.1 1971-1982
Lamborghini Countach & Urraco 1974-1980
Lamborghini Countach & Jalpa 1980-1985
Lancia Fulvia Gold Portfolio 1963-1976
Lancia Stratos 1972-1985
Land Rover Series I 1948-1958
Land Rover Series II & IIa 1958-1971
Land Rover Series III 1971-1985
Land Rover 90 & 110 1983-1989
Lincoln Gold Portfolio 1949-1960
Lincoln Continental 1961-1969
Lincoln Continental 1969-1976
Lotus and Caterham Seven Gold Portfolio 1957-1989
Lotus Cortina Gold Portfolio 1963-1970
Lotus Elan Gold Portfolio 1962-1974
Lotus Elan Collection No.2 1963-1972
Lotus Elite 1957-1964
Lotus Elite & Eclat 1974-1982
Lotus Turbo Esprit 1980-1986
Lotus Europa Gold Portfolio 1966-1975
Marcos Cars 1960-1988
Maserati 1965-1970
Maserati 1970-1975
Mazda RX-7 Collection No.1 1978-1981
Mercedes 190 & 300SL 1954-1963
Mercedes 230/250/280SL 1963-1971
Mercedes Benz SLs & SLCs Gold Portfolio 1971-1989
Mercedes Benz Cars 1949-1954
Mercedes Benz Cars 1954-1957
Mercedes Benz Cars 1957-1961
Mercedes Benz Competion Cars 1950-1957
Mercury Muscle Cars 1966-1971
Metropolitan 1954-1962
MG TC 1945-1949
MG TD 1949-1953
MG TF 1953-1955
MG Cars 1959-1962
MGA & Twin Cam Gold Portfolio 1955-1962
MGB MGC & V8 Gold Portfolio 1962-1980
MGB Roadsters 1962-1980
MGB GT 1965-1980
MG Midget 1961-1980
Mini Cooper Gold Portfolio 1961-1971
Mini Moke 1964-1989
Mini Muscle Cars 1961-1979
Mopar Muscle Cars 1964-1967
Morgan Three-Wheeler Gold Portfolio 1910-1952
Morgan Cars 1960-1970
Morgan Cars Gold Portfolio 1968-1989
Morris Minor Collection No.1
Mustang Muscle Cars 1967-1971
Oldsmobile Automobiles 1955-1963
Old's Cutlass & 4-4-2 1964-1972
Oldsmobile Muscle Cars 1964-1971
Oldsmobile Toronado 1966-1978
Opel GT 1968-1973
Packard Gold Portfolio 1946-1958
Pantera Gold Portfolio 1970-1989
Panther Gold Portfolio 1972-1990
Plymouth Barracuda 1964-1974
Plymouth Muscle Cars 1966-1971
Pontiac Tempest & GTO 1961-1965
Pontiac Firebird and Trans-Am 1973-1981
High Performance Firebirds 1982-1988
Pontiac Fiero 1984-1988
Pontiac Muscle Cars 1966-1972
Porsche 356 1952-1965
Porsche Cars in the 60's
Porsche Cars 1960-1964
Porsche Cars 1964-1968
Porsche Cars 1968-1972
Porsche Cars 1972-1975
Porsche Turbo Collection No.1 1975-1980
Porsche 911 1965-1969
Porsche 911 1970-1972
Porsche 911 1973-1977
Porsche 911 Carrera 1973-1977
Porsche 911 Turbo 1975-1984
Porsche 911 SC 1978-1983
Porsche 914 Gold Portfolio 1969-1976
Porsche 914 Collection No.1 1969-1983
Porsche 924 Gold Portfolio 1975-1988
Porsche 928 1977-1989
Porsche 944 1981-1985
Range Rover Gold Portfolio 1970-1992
Reliant Scimitar 1964-1986
Riley 11/2 & 21/2 Litre Gold Portfolio 1945-1955
Rolls Royce Silver Cloud Gold Portfolio 1955-1965
Rolls Royce Silver Shadow 1965-1981
Rover P4 1949-1959
Rover P4 1955-1964
Rover 3 & 3.5 Litre Gold Portfolio 1958-1973
Rover 2000 + 2200 1963-1977
Rover 3500 1968-1977
Rover 3500 & Vitesse 1976-1986
Saab Sonett Collection No.1 1966-1974
Saab Turbo 1976-1983
Shelby Mustang Muscle Portfolio 1965-1970
Stubebaker Gold Portfolio 1947-1966
Stubebaker Hawks & Larks 1956-1963
Sunbeam Tiger & Alpine Gold Portfolio 1959-1967
Thunderbird 1955-1957
Thunderbird 1958-1963
Thunderbird 1964-1976
Toyota Land Cruiser 1956-1984
Toyota MR2 1984-1988
Triumph 2000. 2.5. 2500 1963-1977
Triumph GT6 1966-1974
Triumph Spitfire Gold Portfolio 1962-1980
Triumph Stag 1970-1980
Triumph Stag Collection No.1 1970-1984
Triumph TR2 & TR3 1952-60
Triumph TR4-TR5-TR250 1961-1968
Triumph TR6 Gold Portfolio 1969-1976
Triumph TR7 & TR8 1975-1982
Triumph Herald 1959-1971
Triumph Vitesse 1962-1971
TVR Gold Portfolio 1959-1990
Valiant 1960-1962
VW Beetle Collection No.1 1970-1982
VW Golf GTi 1976-1986
VW Karmann Ghia 1955-1982
VW Kubelwagen 1940-1975
VW Scirocco 1974-1981
VW Bus. Camper. Van 1954-1967
VW Bus. Camper. Van 1968-1979
VW Bus. Camper. Van 1979-1989
Volvo 120 1956-1970
Volvo 1800 Gold Portfolio 1960-1973

## BROOKLANDS ROAD & TRACK SERIES
Road & Track on Alfa Romeo 1949-1963
Road & Track on Alfa Romeo 1964-1970
Road & Track on Alfa Romeo 1971-1976
Road & Track on Alfa Romeo 1977-1989
Road & Track on Aston Martin 1962-1990
Road & Track on Auburn Cord and Duesenburg 1952-1984
Road & Track on Audi & Auto Union 1952-1980
Road & Track on Audi 1980-1986
Road & Track on Austin Healey 1953-1970
Road & Track on BMW Cars 1966-1974
Road & Track on BMW Cars 1975-1978
Road & Track on BMW Cars 1979-1983
Road & Track on Cobra, Shelby & GT40 1962-1983
Road & Track on Corvette 1953-1967
Road & Track on Corvette 1968-1982
Road & Track on Corvette 1982-1986
Road & Track on Corvette 1986-1990
Road & Track on Datsun Z 1970-1983
Road & Track on Ferrari 1950-1968
Road & Track on Ferrari 1968-1974
Road & Track on Ferrari 1975-1981
Road & Track on Ferrari 1981-1984
Road & Track on Ferrari 1984-1988
Road & Track on Fiat Sports Cars 1968-1987
Road & Track on Jaguar 1950-1960
Road & Track on Jaguar 1961-1968
Road & Track on Jaguar 1968-1974
Road & Track on Jaguar 1974-1982
Road & Track on Jaguar 1983-1989
Road & Track on Lamborghini 1964-1985
Road & Track on Lotus 1972-1981
Road & Track on Maserati 1952-1974
Road & Track on Maserati 1975-1983
Road & Track on Mazda RX7 1978-1986
Road & Track on Mazda RX7 & MX5 Miata 1986-1991
Road & Track on Mercedes 1952-1962
Road & Track on Mercedes 1963-1970
Road & Track on Mercedes 1971-1979
Road & Track on Mercedes 1980-1987
Road & Track on MG Sports Cars 1949-1961
Road & Track on MG Sprots Cars 1962-1980
Road & Track on Mustang 1964-1977
Road & Track on Nissan 300-ZX & Turbo 1984-1989
Road & Track on Peugeot 1955-1986
Road & Track on Pontiac 1960-1983
Road & Track on Porsche 1961-1967
Road & Track on Porsche 1968-1971
Road & Track on Porsche 1972-1975
Road & Track on Porsche 1975-1978
Road & Track on Porsche 1979-1982
Road & Track on Porsche 1982-1985
Road & Track on Porsche 1985-1988
Road & Track on Rolls Royce & B'ley 1950-1965
Road & Track on Rolls Royce & B'ley 1966-1984
Road & Track on Saab 1955-1985
Road & Track on Toyota Sports & GT Cars 1966-1984
Road & Track on Triumph Sports Cars 1953-1967
Road & Track on Triumph Sports Cars 1967-1974
Road & Track on Triumph Sports Cars 1974-1982
Road & Track on Volkswagen 1951-1968
Road & Track on Volkswagen 1968-1978
Road & Track on Volkswagen 1978-1985
Road & Track on Volvo 1957-1974
Road & Track on Volvo 1975-1985
Road & Track - Henry Manney at Large and Abroad

## BROOKLANDS CAR AND DRIVER SERIES
Car and Driver on BMW 1955-1977
Car and Driver on BMW 1977-1985
Car and Driver on Cobra, Shelby & Ford GT 40 1963-1984
Car and Driver on Corvette 1956-1967
Car and Driver on Corvette 1968-1977
Car and Driver on Corvette 1978-1982
Car and Driver on Corvette 1983-1988
Car and Driver on Datsun Z 1600 & 2000 1966-1984
Car and Driver on Ferrari 1955-1962
Car and Driver on Ferrari 1963-1975
Car and Driver on Ferrari 1976-1983
Car and Driver on Mopar 1956-1967
Car and Driver on Mopar 1968-1975
Car and Driver on Mustang 1964-1972
Car and Driver on Pontiac 1961-1975
Car and Driver on Porsche 1955-1962
Car and Driver on Porsche 1963-1970
Car and Driver on Porsche 1970-1976
Car and Driver on Porsche 1977-1981
Car and Driver on Porsche 1982-1986
Car and Driver on Saab 1956-1985
Car and Driver on Volvo 1955-1986

## BROOKLANDS PRACTICAL CLASSICS SERIES
PC on Austin A40 Restoration
PC on Land Rover Restoration
PC on Metalworking in Restoration
PC on Midget/Sprite Restoration
PC on Mini Cooper Restoration
PC on MGB Restoration
PC on Morris Minor Restoration
PC on Sunbeam Rapier Restoration
PC on Triumph Herald/Vitesse
PC on Triumph Spitfire Restoration
PC on VW Beetle Restoration
PC on 1930s Car Restoration

## BROOKLANDS HOT ROD 'MUSCLECAR & HI-PO ENGINE SERIES
Chevy 265 & 283
Chevy 302 & 327
Chevy 348 & 409
Chevy 350 & 400
Chevy 396 & 427
Chevy 454 thru 512
Chrysler Hemi
Chrysler 273, 318, 340 & 360
Chrysler 361, 383, 400, 413, 426, 440
Ford 289, 302, Boss 302 & 351W
Ford 351C & Boss 351
Ford Big Block

## BROOKLANDS MILITARY VEHICLES SERIES
Allied Mil. Vehicles No.1 1942-1945
Allied Mil. Vehicles No.2 1941-1946
Dodge Mil. Vehicles Col. 1 1940-1945
Military Jeeps 1941-1945
Off Road Jeeps 1944-1971
Hail to the Jeep
Complete WW2 Military Jeep Manual
US Military Vehicles 1941-1945
US Army Military Vehicles WW2-TM9-2800

## BROOKLANDS HOT ROD RESTORATION SERIES
Auto Restoration Tips & Techniques
Basic Bodywork Tips & Techniques
Basic Painting Tips & Techniques
Camaro Restoration Tips & Techniques
Chevrolet High Performance Tips & Techniques
Chevy-GMC Pickup Repair
Custom Painting Tips & Techniques
Engine Swapping Tips & Techniques
Ford Pickup Repair
How to Build a Street Rod
Mustang Restoration Tips & Techniques
Performance Tuning - Chevrolets of the '60s
Performance Tuning - Ford of the '60s
Performance Tuning - Mopars of the '60s
Performance Tuning - Pontiacs of the '60s

**BROOKLANDS BOOKS**

## CONTENTS

| Page | Title | Publication | Date | Year |
|---|---|---|---|---|
| 5 | The Series II Land Rover | Motor | April 16 | 1958 |
| 6 | More Refined Land Rover | Autocar | April 18 | 1958 |
| 8 | Australia's Toughest Car Test | Wheels | April | 1958 |
| 12 | Rover in Central Africa | Motor | May 14 | 1958 |
| 14 | New Overhead-Valve Land Rover | Sports Car Illustrated | June | 1958 |
| 15 | The British Land Rover Road Test | Mechanix Illustrated | | 1958 |
| 18 | Workhorse for the Carriage Trade | Car Life | Aug. | 1959 |
| 22 | All it Needed Was a Wash Road Test | Commercial Motor | Sept. 15 | 1961 |
| 28 | Land Rover Station Wagon | Sports Cars Illustrated | March | 1961 |
| 31 | Land Rover 1961 Petrol Models Service Data | Motor Trader | May 31 | 1961 |
| 39 | Land Rover Series II (Diesel) Service Data | Motor Trader | Oct. 1 | 1958 |
| 40 | Land Rover Road Test | Track & Traffic | June | 1961 |
| 42 | Carmichael Redwing Fire Appliance Road Test | Commercial Motor | June | 1962 |
| 46 | Ideal Rally Car? Road Less Impression | Motor Racing | Jan. | 1962 |
| 47 | Horses for Courses Road Test | Motor | March 6 | 1965 |
| 53 | 'Rover' Caravan Test | Motor | July 15 | 1967 |
| 56 | Tough for the Rough | Autocar | Nov. 19 | 1965 |
| 60 | More Powerful Land Rover | Autocar | April 13 | 1967 |
| 62 | Land Rover Six Road Test | Autocar | July 13 | 1967 |
| 67 | Success Story | Autocar | July 25 | 1968 |
| 68 | You Name it . . . and Land Rover can Supply the Version for the Job | Autocar | July 25 | 1968 |
| 70 | 1964 Land Rover Used Car Test | Autocar | July 25 | 1968 |
| 72 | Sandtrekker | Autocar | Jan. 30 | 1969 |
| 76 | O'Kane on Land Rovers | Road & Track | July | 1969 |
| 77 | Twenty-one this Year | Motor | Aug. 9 | 1969 |
| 78 | Up and Over | Motor | Aug. 9 | 1969 |
| 80 | World Vehicle – Land Rover Success Story | Autocar | Nov. 27 | 1969 |
| 82 | Land Rover – 88 Inches of Get There | Popular Imported Cars | Nov. | 1969 |
| 84 | Adventures in a New Guinea Rent-a-Wreck | 4x4 | Summer | 1982 |
| 87 | Land Rover 2.6 Second Hand Buyer's Guide | Overlander | July | 1982 |
| 90 | From Post to Pillar | Overlander | May | 1982 |
| 92 | Land Rover – Transatlantic Specials and Prototypes | Overlander 4x4 | Jan. | 1983 |
| 94 | Land Rover LtWt Goes V8 | All Wheel Drive | Summer | 1984 |
| 96 | Larman's Landie! | 4x4 | March | 1986 |

**BROOKLANDS BOOKS**

*ACKNOWLEDGEMENTS*

When we first started compiling road test books on Land Rover in the early 80's we were forced to include more than one series of vehicle in each book, due to the paucity of material. Happily, with the great explosion of interest in 4x4s and in Land Rovers in particular during the intervening years, we have now managed to gather enough stories to cover the Series I, II & IIa, III and the 90's and 110's individually. Details of these books will be found on the inside back cover.

We could not have achieved this greater coverage without the encouragement and help of long-standing Land Rover devotees, such as David Shephard who supplied photographs when they were needed. David Gerrard of Liverpool was kind enough to loan his whole collection of 4x4 articles so that we could bring a new dimension to the books by including stories from more diverse sources. David Bowyer who runs the popular Off-Road Centre In Devon was also generous with his help.

Land Rover enthusiasts will I am sure wish to join with us in expressing our thanks to the publishers of All Wheel Drive, Autocar, Car Life, Commercial Motor, 4x4, Mechanix Illustrated, Motor, Motor Racing, Motor Trader, Overlander, Overlander 4x4, Popular Imported Cars, Road & Track, Sports Cars Illustrated, Track & Traffic and Wheels, for generously allowing us to include their informative copyright stories.

R. M. Clarke

1958 CARS

# The Series II Land-Rover

**Many Useful Modifications to the 88-in. Wheelbase Model and a New 2¼-litre Petrol Engine for 109-in. Models**

*A smoother exterior appearance and useful scuttle vents: Series II features seen here on the 88-in. wheelbase model.*

TEN years ago the first Land-Rover was introduced, and since then more than 200,000 have gone into service all over the world. The new Series II Land-Rovers have been developed in the light of the experience thus gained, but the basic design has proved so sound that major changes have not been necessary.

The appearance of the 88-in. wheelbase model has been improved by giving it a less square-cut look, and the wings now have a smooth curve which is carried along the waistline for the full length of the vehicle. Most of the alterations to the technical specification have been introduced to improve the ride, and include minor alterations to the spring rates and redesigned telescopic dampers. The turning circle has been reduced by three feet by increasing the track by 1½ in., and the rear axle is now of the fully-floating type.

Deeper and softer internally-sprung seats and modifications to the angle of the back rests also help to improve the ride for the occupants. Pendant clutch and brake pedals mounted high up on the engine side of the front bulkhead have been adopted to improve the sealing of the front compartment against dust and sand and also to raise the associated control rods to a height where they are unlikely to be blocked by flying stones. Visibility has also been improved by making all windows of non-scratch glass instead of plastics material.

Minor but very practical modifications include the transfer of the fuel filler with its telescopic filter tube from its former position under the driver's cushion to an external location on the right hand side of the body, the introduction of a single bonnet release catch operated by a trigger in the front grille and ingenious new quick-release tailboard catches.

All these improvements are also incorporated in the 109-in. wheelbase model, but the increase in track of 1½ in. has reduced the turning circle by no less than five feet. In addition, the rear cab window has been increased in size and rounded quarter lights fitted to improve visibility. Progressive rate springs have been adopted which provide two inches more travel and are mounted on hangers brought forward to give greater stability.

The new 2¼-litre petrol engine differs from other current Rover engines in that all the valves are in the head instead of only the inlet valves. The valves are push-rod operated by roller-type tappets, the rollers of which follow the cam in lead tin-bronze shoes which in turn slide in cast-iron tappet guides. The three-bearing counterbalanced crankshaft runs in copper-lead bearings with a tin overlay. The new engine is oversquare and has a bore and stroke of 90.47 x 88.9 mm., giving a capacity of 2,286 c.c., and develops 77 b.h.p. at 4,250 r.p.m. It has much in common with the 2-litre Diesel engine, which remains available in both the Series II models.

The station wagon model on the 88-in. chassis is temporarily out of production and the 107-in. wheelbase station wagon is continued in its Series I form.

Prices of the 88-in. model rise by £10, but that of the L.W.B. station wagon is unchanged. The 109-in. prices are: Basic models, £730 (petrol) and £820 (Diesel); the respective de luxe versions cost £20 more. The Price Table will be amended next week.

*77 b.h.p. is developed by the new oversquare 2¼-litre engine which has full o.h.v. operated by roller tappets, push-rods and rockers, many details of this sturdy engine deriving from the 2-litre Diesel.*

# MORE REFINED LAND-ROVER

*Left: In the Series II Land-Rover the sides of the bodywork have been rounded, and better visibility provided for the driver in the cab. Above: Each valve port has its separate outlet in the cylinder head, and the downdraught carburettor is silenced by a large cylindrical air cleaner*

## New 2¼-litre Engine for Series II Long-wheel-based Models

EVERY so often automobile manufacturers produce a design which is a winner from the start—the basic conception is right, and though detail alterations may be called for, in general the vehicle is exactly what is required. It is easier to reach this high standard with a simple layout than with an elaborate specification—as in a very fast car, for instance—and when the problem involves a specialized type of vehicle, the complications are correspondingly greater.

During the last war the American-designed Jeep was manufactured and used in great numbers, and new and reconditioned models came on to the civilian market later as people realized the worth of this type of vehicle. Its ability to deal with large loads and its cross-country capabilities made it invaluable, but it was stark and uncomfortable, and spares became increasingly difficult to obtain.

In 1948 the Rover Company, who had studied the problem and markets methodically, produced the first Land-Rover—a name which very shortly was to become a household word. It was a new departure for peacetime production and the works were geared for an output of 50 vehicles a week. Wheelbase was 6ft 8in and the overall length 11ft; the four-cylinder 1.6-litre engine was similar to that used in the Rover 60 saloon of that time. The transmission included a four-speed gear box with synchromesh on third and top, a low ratio transfer box and four-wheel drive. The body was extremely practical, and built for hard work with a minimum of maintenance.

In a few weeks the demand exceeded the supply, production was stepped up soon to reach 200 units each week, and by 1950 Land-Rovers had brought in £5 million in foreign currency. The original design continued to be modified and improved, as in 1951 when the engine capacity was increased to 2-litres.

Up to the present, more than 200,000 Land-Rovers have been built, and 75 per cent of current production is going overseas. There are five basic models in the range, including the Series I 107in wheelbase station wagon which has the 2-litre petrol engine. Series II includes the 88in wheelbase version with a choice of petrol or diesel engines, and the 109in wheelbase car with either the new 2¼-litre petrol engine or the 2-litre diesel

*Pick-up version of the Land-Rover with chain-suspended tailboard*

## SPECIFICATION

**ENGINE:** No. of cylinders, 4 in line; Bore and stroke, 90.47 x 88.9 mm (3.562 x 3.50in). Displacement, 2,286 c.c. (139.5 cu in); Valve position, Overhead; Compression ratio, 7.0 to 1; Max. b.h.p. (gross), 77 at 4,250 r.p.m.; Max. b.h.p. (nett), 70 at 4,250 r.p.m.; Max. b.m.e.p., 134 lb sq in at 2,500 r.p.m ; Max. torque, 124 lb ft at 2,500 r.p.m. installed; Carburettor, Solex 40 PA10 down-draught; Fuel pump, AC mechanical; Tank capacity, 10 Imp. galls (45 litres); Sump capacity, 11 pints (6 litres); Oil filter, External full flow; Cooling system, Pump, fan and thermostat, pressurized; Battery, 12 v 51 amp-hr.

**TRANSMISSION:** Clutch, Single dry plate, 9in dia, hydraulic operation; Transfer gear box ratios, High, 1.148 to 1; Low, 2.888 to 1; No. oı speeds, main gear box, Four forward; one reverse; Gear lever position, Central, floor mounted; Synchromesh on third and top; No. of speeds, transfer box, Two speed reduction on main gear box output; Overall ratios, High: Top, 5.396; 3rd, 7 435; 2nd, 11.026; 1st, 16.171; Reverse, 13.745 to 1; Low: Top, 13.578; 3rd, 18.707; 2nd, 27.742; 1st, 40.688; Reverse, 34.585 to 1; Final drive, front and rear, Spiral bevel, 4.7 to 1.

**CHASSIS:** Brakes, Girling, two leading shoe, front; leading and trailing shoe, rear; Drum size, 11in. x 2¼in.; Suspension, Beam axle, half-elliptic leaf springs, front and rear; Spring dampers, Telescopic; Wheels, Steel disc, 5-stud fixing; Tyre size, 6.00 x 16 (according to specification); Steering, Recirculating ball; Steering wheel, 17in; No. of turns, 3¾.

**DIMENSIONS:** Long wheelbase, 9ft 1in (277 cm); Track, Front, 4ft 3½in (131 cm), Rear, 4ft 3½in (131 cm); Overall length, 14ft 7in (444 cm); Overall width, 5ft 4in (163 cm); Overall height, 6ft 9in (206 cm); Ground clearance, 9¾in (25 cm); Turning circle, 45ft (13.72 m); Kerb weight, 29½ cwt (1500 kg).

**PERFORMANCE DATA:** Top gear m.p.h. at 1,000 r.p.m. with 7.50 x 16 tyres, 16.9; Torque lb ft per cu in engine capacity, 0.89 (gross); Weight distribution dry, Front, 54 per cent; Rear, 46 per cent.

# More Refined Land-Rover . . .

power unit. De luxe versions of the 109in model are also available.

Any of the production models can, of course, be fitted with a great variety of specialized equipment, and power take-off drives can be applied to the front, centre or rear of the chassis. Radio and heater can be fitted, as with a normal saloon.

A newly introduced 2¼-litre petrol engine, based on the recent 2-litre diesel unit, develops 25 b.h.p. more than the c.i. engine and 2-litre petrol power unit. The crankcase is an iron casting with liberal water passages between each cylinder bore. Cylinder liners are not used. A three-bearing crankshaft is carried in copper lead bearings which have a tin overlay. The front and centre bearings are 2½in dia by 1$\frac{1}{16}$in wide, and the rear main is 1⅝in wide. The connecting rods, which have the big ends split horizontally, can be withdrawn through the cylinder bores.

The valve seats are cut direct in the cylinder head—which is an iron casting—and there is a separate port for each valve. A thermostatically controlled hotspot is formed between the inlet and exhaust manifolds. In determining the gas flow of the new cylinder head, Rover engineers had the collaboration of Mr. Harry Weslake. The previous overhead inlet side exhaust arrangement has been abandoned; all valves are now in the head. The camshaft is placed high up in the right side of the cylinder block, and roller-type tappets are used. A double roller chain which has a hydraulic tensioner drives the camshaft. Two compression and one oil control ring are fitted to the light-alloy pistons, which have fully floating gudgeon pins.

No change has been made in the main gear box and transfer box; the driver has the choice of eight forward and two reverse speeds, and four-wheel drive can be used with either the high- or low-transfer box ratios employed. The hand brake continues to operate on the transmission behind the gear box.

Chassis detail changes include increasing the track by 1½in, which has decreased the turning circle by 3ft for the 88in and 5ft for the 109in wheelbase model. The rear hubs are now fully floating on taper roller bearings. Rear spring brackets are mounted on the outside of the chassis frame and the progressive rate springs have 2in greater travel than in the Series I version. Pendant brake and clutch pedals are fitted, and the fuel tank filler, which has a telescopic neck, is accessible from outside the body.

Increased comfort for the crew is provided by improved, sprung seat cushions and backrests. The de luxe model 109in Land-Rover has an adjustable driving seat, the door inner panels are trimmed, and floor carpet can be supplied. Perspex is replaced by glass in the side windows, and better driving vision is provided by the cab rear window and quarter lights. Externally the rather square look has been softened to some extent by introducing curves into the body and cab lines, without detracting from the functional aspect of the design.

*Pendant pedals provide greater protection for the operating mechanism and conform to the modern tendency of controls. The doors are held open by a spring-cushion arrangement and there is ample room for small objects on the shelves adjacent to the instrument panel*

**New Land-Rover Series II prices are:**

88in Wheelbase Regular with 2-litre petrol engine, **£640.**
88in Wheelbase Regular with diesel engine, **£740.**
109in Wheelbase Long Basic with 2¼-litre petrol engine, **£730.**
109in Wheelbase Long Basic with diesel engine, **£820.**
109in Wheelbase Long De Luxe with 2¼-litre petrol engine, **£750.**
109in Wheelbase Long De Luxe with diesel engine, **£840.**
107in Wheelbase Long Land-Rover Station Wagon, **£815** basic, £408 17s purchase tax, **£1,223 17s** total.

*ASSOCIATION OF ROVER CLUBS LTD.*

*Patrons:*
Tom Barton, O.B.E.
Major B. Hervey-Bathurst, O.B.E.

*Reg. Office and Hon. Secretary:*
G.R. Day, 10 Highfield Road, Bagslate, Rochdale, OL11 5RZ
Telephone: Rochdale (0706) 30200

**Land Rover Series Two Club**
*Secretary*: Ross Floyd, 2 Brockley End Cottages, Cleeve, Avon. BS19 4PP.
Tel.: (027 583) 3772 or (0272) 669893.

**The Land Rover 2 Litre, Series II Register**
*Secretary*: Eric Cowell, Breeds Farm, 57 High Street, Wicken, Ely, Cambs. CB7 5XR.
Tel.: (0353) 720309.

# AUSTRALIA'S TOUGHEST

... an ALAN GIBBONS feature

*Into this mountain torrent went our Rover, with the icy water bonnet-high in the middle. Current was so strong that the Land Rover could be felt slipping sideways — but the engine never missed a beat!*

*Second highest town in Australia is Cabramurra, where "Wheels" sent Gibbons to conduct this test. The vehicle park at foreground houses over 350 vehicles — about half of which are Land Rovers.*

THIS can fairly be called a road test of a Land Rover — yet it was one occasion when stop watches, Tapley meters, and slide rules were all deliberately left home at the office!

It is a practical test; in a vehicle which lays no claim to fast quarter miles, to 60 m.p.h. in any record number of seconds, or to any fantastic petrol consumption. This vehicle was built simply for a lifetime of hard work under atrocious conditions — and that's exactly how we tested it!

"Wheels" obtained the approval of the Snowy Mountains Authority to make the test, and Grenville Motors, of Sydney, assisted us make all arrangements. But when we left Cooma, we still did not know from which

# CAR TEST

*Where, we ask you, can one test a Land Rover adequately near Sydney? We thought we'd do it properly; packed Gibbons off to the Snowy Mountains Hydro to see what these fabulous little waggons are doing in Australia's toughest going.*

*Dip your fibreglass lid to boss Roverman Bill Shaw! Bill, who supervises more than 185 Land Rovers, has been seven years on the Snowy; is recognised as probably the best driver in the area. In snow, blizzard, heat, dust and accident, Bill sees that his Land Rovers stay on the job; in his spare time he went along to make sure Gibbons didn't kill himself.*

*Rover in the rain! Following behind a grader, it was necessary to cross this boggy bank several times — sheer simplicity to a Land Rover*

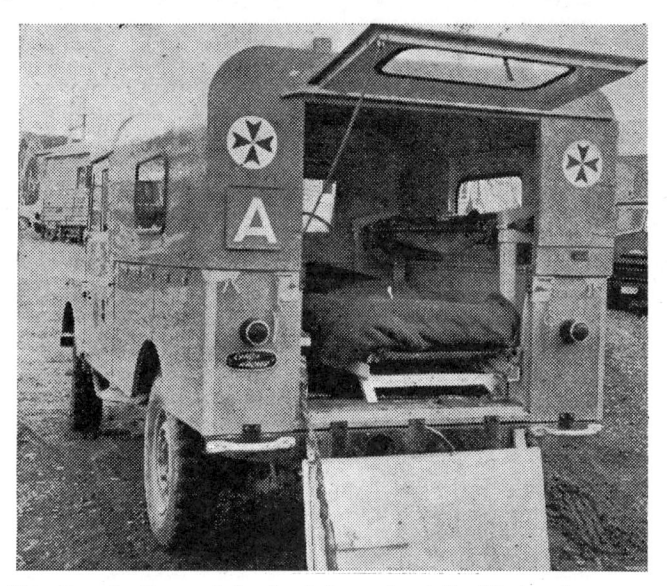

*Ubiquitous is the word for Land Rover. Specially fitted ambulances, each with their own doctor, are always on call at the Snowy; have done a mighty job in "impossible" country.*

transport park we would finally requisition a Land Rover!

We drove 70 miles from Cooma to the mountain village of Cabramurra, where more than 185 Land Rovers are permanently stationed, and here we met Bill Shaw — the regional transport officer — who is responsible for all vehicles in the area.

Bill has been on the Snowy Scheme for seven years, and he turned out to be a tower of strength in every way.

He is a man who knows every track and every by-road throughout the length and breadth of the Australian Alps, and who has driven Land Rovers over this shocking terrain in all weather conditions. He happily rides his Rover when it is slithering down a snow-covered mountain face;

he is accustomed to belting it through raging mountain streams; while frequently he is called upon to make a dash in a Land Rover ambulance to bring out an injured worker. We could say that Shaw's life revolves around Rovers.

Before we left the transport office, Shaw tossed me a lightweight fibreglass topee — a lifesaving precaution worn by every person who enters the working areas of this great scheme.

"While we're in the Rover," explained Shaw, "it is essential for you to wear that helmet. They have been responsible for saving many lives when vehicles have been struck by falling rocks. It might be a little uncomfortable, but keep the damn thing on your head!"

### Random Choice...

In the transport park there were more than 60 of the 185 Land Rovers, all parked in military fashion and carrying a distinguishing signplate, indicating which department employed that particular vehicle.

I selected a Rover at random — one which they told me belonged to the regional engineer, and which was used for six days every week. Its speedo reading was 6274 miles, and, according to its log book, it had been serviced three days before our visit.

Outwardly it appeared the same as any Rover that might be seen in any country district. It had the standard bar-tread tyres and a metal canopy, while the spare wheel was mounted in its conventional position, i.e., on

"This," said the man, "I call Land Rover country!" Gradients of the order of 1 in 3, loose shifting surfaces, ice, snow, and rushing water are all in the day's work to a Snowy Scheme Land Rover. This "road" climbs a sheer 3000 feet.

top of the bonnet. However, closer examination showed a few extras which had been fitted to comply with the local mountain conditions. Fog lamps were mounted between the grille and the bumper bar, while a Lucas "flame thrower" and radio aerial were fitted to the top of the canopy.

Inside the cabin there were also a few minor differences. A two-way radio telephone had been installed on the left hand side of the dashboard, and two de-misters were in place to help assist vision in near freezing temperatures.

Otherwise, the vehicle was a standard Model 88 Land Rover in every way.

The motor fired instantly, and we waited for a few minutes for it to warm up before edging the vehicle out of the car park. Our first run was towards what is known as T.1, where hundreds of men are working on a huge tunnel. The road to the lookout, the last point where members of the general public can see this part of the scheme, was comparatively good.

Although five inches of rain had fallen at Cabramurra during the previous 24 hours, the road's surface was quite firm, and the Rover plodded along at a steady 35-40 m.p.h.

### Straight Down . . .

But once past this lookout point the road deteriorated rapidly! Leaving behind a signpost which read: "The Public Must Not Proceed Beyond This Point", the road unwound like a huge ribbon and threaded its way straight down the mountain side. It dropped more than 3,000 feet — and there was no safety fence or even any retaining stones to check a vehicle should it get out of control!

For most of the time the road was less than 20 feet in width — yet giant 15 ton diesel trucks were making the trip up and down continuously, day and night.

The longest straight on this road was a little over 200 yards long, and many of the hair-pin bends required drivers of heavy vehicles to stop and reverse before they could proceed. On the inside of the track I noticed a three foot deep gully, while on the outer edge guide posts standing more than nine feet high were placed every hundred yards or so.

Shaw explained that the guide posts were for winter time, when up to six feet of snow covered the roadway. (Snow plough operators use these posts as their guide when clearing the track.)

That inside gutter, too, has saved many lives. Drivers who sense they are heading for trouble edge the vehicle into this depression; thus keeping well clear of the sheer drop beyond the outside edge of the road.

"Every driver on the Snowy keeps to the right hand side of the roadway on a descent," Shaw explained. "They realise that the outside edge spells death; while if they run into the inside gutter they stand a good chance of getting out alive."

### Heavy Traffic . . .

Widened sections of the track permit two vehicles to pass, and rarely does a vehicle complete the trick descent or ascent without having to pull into one or more of these sections, so heavy is the traffic.

Towards the bottom of the mountainous roadway, the surface became extremely treacherous. The heavy vehicles had churned the top soil into a sea of slippery mud up to 9" deep, and we were forced to use low gear. The Rover scrambled through this hazardous part without any undue slipping or sliding; with a continuous thud-thud-thud coming from beneath the floor as dollops of mud chucked up from the front wheels thumped underneath the body.

At the bottom of T.1, we pointed the Rover into a narrow but fast-flowing creek. The bonnet promptly disappeared beneath the rushing water, and the tail of the vehicle could be felt swaying in the torrent. Steadily the Land Rover lumbered through, its motor never missing a beat, and within a matter of minutes we were back on dry land and bouncing over a rocky outcrop which adjoined the stream.

"Let's go back to Cabramurra via the Old Road," Shaw explained, "and then you'll see how a Rover works in what I call true Rover country."

The Old Road turned out to be nothing more than a billygoat track! Seldom more than 10 or 12 feet wide, it was the original track used before the main roadway was hacked out of the mountainside. It is now an emergency road in case the main thoroughfare is ever cut by landslide or through accident.

Very loose, the surface was mostly chunks of rock as big as a man's fist. Deep 18" gullies crossed it higgledy piggledy, cut by rainwater rushing down the mountainside. Hair-pin bends were so severe that even the sawn-off Land Rover was forced to stop and reverse on some of them. On numerous occasions we were in low gear, but generally speaking the ascent was completed in second gear — with four-wheel drive engaged constantly, of course.

It was over tracks such as this that the excellence of the Rover's suspension were appreciated to the full. Naturally we received a rough ride — but in any other vehicle it would have been back-breaking.

### Sheer Drop . . .

On one sharp bend the wheels of the Rover were poised on the brink of a sheer drop of thousands of feet; but this meant nothing to a driver such as Bill Shaw, who had taken the wheel. Knowing every inch of the road, he knew the exact spot where it was necessary to swing the wheel in order to complete each bend in one sweep.

"We give every driver who comes to the Snowy a thorough test before we let him take charge of a vehicle," explained our guide. "Then he is licensed to drive only a certain type of machine. If he wants to earn more money he must qualify for a heavier vehicle, after first completing another driving course. It's the only way out, really, because we can only afford to have fully qualified and really capable specialists at the wheel of a vehicle in this type of country."

With the motor roaring, and in first gear, we slowly but inexorably dragged our way up the emergency road; yet at the top of this long, slow, three mile pull, over gradients as high as 1 in 3 in places, the Rover's motor showed no signs of boiling.

# AUSTRALIA'S TOUGHEST CAR TEST

Next I decided to aim the Land Rover along a newly formed roadway on which a grader had been at work immediately prior to the falling of five inches of rain! (Anyone care to follow in a car?)

This roadway ran along through miles of gooey black soil country which was particularly slippery, and in places, boggy.

I found while scooting along some of the harder, newly graded sections, that the Rover was inclined to wag its tail, but it was never difficult to maintain under perfect control. After we had traversed the newly made surface, however, there was nothing else for it but to engage four wheel drive. Now we were lobbing over ditches up to three feet in depth; so soft that our wheels sank in nearly to hubcap depth. It was in country such as this that the suspension came really into its own, and at times the vehicle was on such an angle that there would be more than 12 inches showing between the top of a tyre to the bottom edge of the mudguard!

## Bitumen...

For the next couple of hours, for a change, we motored along main roads. On bitumen surfaces I found that the Rover behaved in much the same manner as any sedan motor car. The all-steel canopy with its sliding windows was quite draught-proof, and the absence of body rattles was commendable. At 45 m.p.h. (the maximum speed permitted by the Snowy Authority), the steering was light and precise, while the passenger comfort was first class. During this period of normal driving we tested the Rover's built-in heater.

After a few minutes there was so much hot air inside the cabin that we were forced to turn it off again!

"They make life bearable in winter, though," said Shaw. "I spend four to five hours every day in my Rover, and when the temperature is down near freezing point you really need that heater, boy! And de-misters, too, are essential."

Our final gambit included another mountain descent, and then up again to Tumut Ponds, the site of another gigantic dam project.

## More Mud...

For many miles we belted along a reasonably flat byway which ran through a loam type of soil. Here the going was good, except for a few bends where black squelching mud caused a bit of tail slide.

On our way we skittered through numerous miniature lakes — many up to 200 yards long and moree than 12 inches deep. Although we were travelling at up to maximum allowable speed, the motor never faltered, and, later on, examination showed it to be perfectly dry!

The roadway down the mountainside at Tumut Ponds was worse, believe it or not, than that leading to the T.1 shaft. The oozing mud was more than a foot deep in places, and when a drop of up to 2,000 feet loomed ahead and the Rover began to slide, willynilly, I simply grabbed for the panic bar and hung on. But Shaw, without blinking an eyelid, casually swung the wheel, and time after time we would miss the road's edge by inches. Again on this road I noticed the deep lifesaving gutter which ran the full distance along the inside edge.

Down towards the bottom there was a patch of mud which had recently been churned up by two bulldozers; where the metal tracks of the caterpillar machinery had dug down close to two feet into the mud, and the mess resembled the now famous bog which I remember encountering north of Kingoonya during the now-notorious 1956 Ampol Trial.

Through this quarter mile of horror Shaw pushed the Rover steadily in four-wheel drive. Gripping the slimy surface, the bar-tread tyres dug deep into the softness and we emerged with the little car plastered with mud from base to breakfast.

To the Rover this was simply all in the day's work!

Another back road taken while returning to Cabramurra proved to be just as hair-raising as that on the ascent from T.1. Tight hairpins, stony, dangerous country which could not have been negotiated by any two-wheel drive vehicle, and 1 in 4 gradients were encountered throughout our four mile trip back to the regional car park.

Once there the radiator was checked, and, although much of the journey had been completed in four-wheel drive, there was still no trace of boiling!

"You ought to see what these vehicles go through in wintertime," Bill Shaw said, patting the bonnet fondly. "For eight months of the year they are travelling through snow and slush constantly, and four-wheel drive is never disengaged. Even so, the motors still stand up for more than 30,000 miles without attention. This would probably equal all of 90,000 miles of normal country use.

"When a motor shows signs of wear we bung a new one in, and it costs us a little over £105. One of the reasons why these vehicles have been so successful in the Snowy is because they are regularly serviced and driven by qualified drivers," he added. "With ordinary drivers at the wheel I hate to think what might happen.

"It would be horrible!

"A daily report is completed by each driver, and if he recommends that a certain job be done, the vehicle is taken out of service immediately and repaired," he pointed out.

Workmen on the Snowy Scheme rely implicitly on Land Rovers to get to and from work, for supplies, and as a method of transporting injured workers (there are many) to a point where the conventional ambulance can take over the job of medical transportation.

There were three Land Rovers at the Cabramurra transport park, while I was there, which had been specially converted for ambulance work. In emergency, a doctor travels with the Rover driver out to the scene of the accident, and he renders first aid during the return trip. Each "meat waggon" has a full length stretcher fitted, and carries a supply of drugs and medical equipment.

Rovers have also been used to erect power lines through virgin country where no other vehicle can travel. Surveyors and geologists use them exclusively too when making a reconnoitre of the rugged countryside.

It was a long, rugged haul up to the Snowy to write this story for "Wheels", and I'm afraid my own car (left at Cabramurra) has never been quite the same since. But it was worth it! I know now why Sir William Hudson, Commissioner for the Scheme, once stated:—

"Without Land Rovers we would never have been able to keep to schedule on this project. Instead, we are now months ahead on every working!"

## Technical Details...

MAKE: Two-door Model 88 Land Rover — 6/7 passengers and fitted with all-metal turret. Test vehicle from the Snowy Mountains Authority; arranged by courtesy of Grenville Motors Pty Ltd., Sydney.
PRICE and AVAILABILITY: £1,160, excluding sales tax. (Not applicable for farm use). Top £74 extra. Delivery—Minimum 6 weeks.
ENGINE: 4 cylinder. Bore and stroke, 77.8 x 105 mm. Capacity, 1997 c.c.; compression ratio, 6.8 to 1. RAC rating, 15 h.p.; developing 52 b.h.p. at 4,000 r.p.m. Carburettor, single Zenith downdraft. Capacities: radiator, 17 pints; sump, ?? pints; fuel tank, ?? gallons.
TRANSMISSION: Clutch, single dry plate with cushioned drive. Gearbox fitted with four-speeds and reverse with s/m. in top and third. Two-speed transfer box in main gearbox output. (High 1.148 to 1 — low 2.888 to 1.)
RATIOS: Main Gearbox — Top: Transfer box, high ratio, 5.396; low ratio, 13.578; 3rd, 7.435, 18.707; 2nd, 11.026; 27.742; 1st, 16.171, 40.688; reverse, 13.745, 34.585.

### All-Victoria Rally

The BP All-Victoria Rally, finishing in Melbourne, has to date attracted over 50 entrants to the start on Thursday, March 13. Starting points are at Sydney, Melbourne, and Adelaide; and all contestants will assemble at Deniliquin, following which a tight time schedule will be adhered to for the 1,500 mile Rally section to Melbourne.

Prizemoney at time of writing totals £1,250.

Many well-known trials and racing personalities will be taking part, and much interest is centring on the team of Lloyd-Hartnetts which have entered from Sydney—this being the first appearance of these little cars in a competitive event in Australia.

PIPELINES to PROSPERITY

A Land-Rover meets the local transport on a journey from Central Africa to the Union.

# Rover in Central Africa

THE tremendous African Continent has always presented a challenge to the British industrialist. Stretching through two hemispheres, the Continent is too large to consider easily as a whole, and even Central Africa staggers the imagination with its diversity of climate, temperature and conditions, peoples and civilizations. An awakened giant, this central region, once the home of pastoral nomads, unexplored and undeveloped, is now expanding and making immense strides towards what shows every signs of being a prosperous future.

One should remember that until the 19th century only the coastal regions had any contact with the world outside Africa, while inland the natural barriers, particularly distance, kept Central Africa undisturbed and undeveloped. The great explorers of the 19th century opened up the interior, but even then the area lay with her secrets revealed but unconquered. The development of the land through agriculture, and the utilization of the mineral wealth were begun, and the land began to show signs of a promising future. The commerce and industry thus started multiplying until the whole new world began to develop and pour forth its riches, not freely but as a reward for hard labour and perseverance, allied to skill and ingenuity.

In developing this new wealth and means of trade, the great natural barrier of distance made transport a vital factor in the success or failure of the enterprise involved. Transport was needed to carry food and supplies to the interior and to carry the results of the enterprise from the territories to the coast where they could be sent to awaiting markets. The ability to conquer vast distances was a pre-requisite of the successful development of the land.

By

G. LLOYD

DIXON

*Executive Director (Sales), The Rover Co. Ltd.*

*Servicing conditions in the African interior can often be extremely primitive. This Land-Rover is raised for servicing by driving up the planks onto the rear barrels, then over-balancing — see-saw fashion — onto the front ones.*

While Europe, with its densely populated areas, has always had the labour and wealth to meet her needs for transport as these needs developed, Central Africa's small population faced with vast distances between centres of civilization has been forced to make the development of roads await their turn in the general pattern of development until labour and materials were available to create a wide network.

To support the ever-growing population by the continual creation of additional wealth, Central Africa must develop agriculture and exploit its mineral wealth for the creation of new homes and new industries. Mobility is the key factor of this development. Not mobility over smooth tarmacadam roads, but free and rapid mobility where no roads exist. As man's ingenuity has invariably provided the answer to man's need, the vehicle with power to all four wheels, such as the Land-Rover, is providing the answer to the needs for mobility in Central Africa. Right from the start of Land-Rover manufacture in 1948 it was evident that the Land-Rover fulfilled a world need, particularly in territories such as Central Africa. It also became evident that not only manufacturing programmes had to be enlarged, and factory capacity increased, but also additional versions of this specialized vehicle would be needed to meet special requirements.

The Rover Company now produce four separate versions of this 4-wheel drive vehicle, and all are designed and robustly manufactured to operate and continue operating in all conditions of climate and terrain. In Central Africa, as elsewhere, the Land-Rover is regarded as more than just a vehicle, it is the answer to the need for mobility under all conditions. From Mombasa on the coast of East Africa to Ndola in Northern Rhodesia, from Beira in Portuguese East Africa to Salisbury and Bulawayo in Southern Rhodesia, the Land-Rover is regarded as reliable transport, giving the inhabitants mobility anywhere, any time and under any conditions. It is significant that 75% of the Rover Company's Land-Rover production is exported, and the expanding markets of Central Africa eagerly take up 13% of this export total through a chain of distributors and dealers who are equipped to give the spares and services facilities that this important market warrants. Ten years of Land-Rover development, concentrated engineering and careful application of lessons learned in all countries, under all conditions, have improved the breed. The vastness of Central Africa, and remoteness and inaccessibility of certain areas, present tremendous difficulties in ensuring immediate availability of spares and service when required, and users make their Land-Rovers work fantastically hard for their living.

Distributors of the Rover Company's products meet the challenge involved in keeping Land-Rovers always up to the mark by holding considerable stocks at all main centres, and stocks of certain fast moving items in remote areas. These stocks of spares and consumable items represent considerable export business for the United Kingdom, the take-up of spares in East Africa alone being in the region of £100,000 per year, and the holding of such spares is an expression on the part of the Rover Company's distributors of their confidence in the future of Central Africa. Amongst the population of Central Africa there exists a genuine desire to build and develop their new world. Though cities have been built comparable to those in Western Europe, the roads built between them are not as we in Europe know them, but vary from smooth tarmacadam to corrugated graded earth. On these roads cars are required to operate for the transport in comfort of men and their families; for the ability to operate in these conditions in heat, mud, sand, dust and humidity, and to take the tremendous pounding given by these roads, an especially high standard of car is required. Many British manufacturers are producing cars to meet these standards and amongst them the Rover Company has a range of vehicles which, by their inherent design and robust construction, provide the answer to family and personal transport by giving luxurious trouble-free motoring with a European standard of comfort.

The importance to Britain's export drive in the Central African market must not be under-estimated as these markets must, and inevitably will, expand. British motor manufacturers can constantly assist this expansion by providing vehicles built to meet and conquer African conditions in the knowledge that, dependent on our success, we shall continue jointly to enjoy the fruits of expansion and development in this new world.

# NEW OVERHEAD-VALVE LAND ROVER

**More powerful 2¼-litre engine. Face-lift for 2-litre short wheelbase model. Improved suspension**

*Curved quarter rear windows provide excellent visibility on the new 2¼-litre, 109-in. wheelbase, 4-wheel-drive Land Rover. The new smooth body lines are apparent. The 2-litre Rover Diesel engine is an optional power-unit*

AFTER ten years and a production of 200,000, that most sporting of non-sports cars, the Land Rover has undergone a face-lift and several technical improvements to "Live in peaceful co-existence". The quotation is from Rover joint managing director M. C. Wilks' speech at the Land Rover Tenth Birthday and demonstrations on 15 April.

Since its original introduction in 1948 (as a 1½-litre) the Land Rover has distinguished itself in trips of exploration to jungle, desert and remote outposts, it has had the privilege of conveying members of the Royal Family on many occasions, and has become synonymous with modern farming, agriculture, fire-fighting, civil engineering and many other duties.

Later models were powered by a 2-litre engine (based on the Rover 60 saloon model unit), then long wheelbase and Diesel models were evolved and the Land Rover became Britain's most exported commercial vehicle. Its reliability and rugged construction and its consistent appeal to operators in the world's undeveloped territories has led to more than 75 per cent of production being directed to 150 different countries.

The entirely new 109 in. wheelbase model has a more sleek appearance obtained by giving the wings a smooth curve which continues along the waistline for the entire length of the vehicle. Seating is improved by the use of deeper and softer cushions and squabs. Softer road springs and greater suspension movement also contribute to improved comfort, beside ensuring faster cross-country progress. Pendant type clutch and brake pedals are now fitted, high on the engine side of the bulkhead.

Perhaps the most interesting feature of the new Land Rover is its "petrol" power-unit. To ensure rigidity and consequent reliability, the Rover Company in its customary thorough manner has evolved a new 2¼-litre engine by utilising the basis of their Diesel engine crankcase. When it is mentioned that the oil engine functions on a compression ratio of 19½ to 1, the stiffness of the casting can be imagined.

Unlike the saloon (and the 88 in. wheelbase Land Rover) models, the new vehicle has an all-overhead-valve layout which permits the use of an "over-square" bore/stroke ratio with consequent large valves, good combustion chamber filling and excellent torque. Comparisons of the

*The 88 in. wheelbase Land Rover also has a "new look", redesigned dampers, seats and back rests. This model retains the 2-litre i.o.e. petrol engine, but a 2-litre Diesel unit is also available*

old long-stroke 2-litre and new short-stroke 2¼-litre make interesting reading. The 2-litre unit (bore 77.8 mm. by 105 mm.) develops 52 b.h.p. at 4,000 r.p.m. on a compression ratio of 6.9 to 1. Maximum torque is 101 lb. ft. at 1,500 r.p.m.

The new 2¼-litre unit develops 77 b.h.p. at 4,250 r.p.m. on a compression ratio of 7 to 1. Maximum torque is 124 lb. ft. at 2,500 r.p.m. A useful engine with just the type of performance for the job in hand. Its crankcase and bearings designed for such high pressures should have an indefinite life with a compression ratio of only 7 to 1. The 2-litre Diesel engine is also available in the new model which has an inch greater ground clearance and is 1½ in. longer in the wheelbase than the Series I. It is also 1 7/16 in. wider and 3 in. lower.

The Series II 88 in. wheelbase model has also been given the new "smooth look", and the hood design has been modified to give a neater and more graceful appearance. It is available with 2-litre petrol or Diesel engine. Redesigned telescopic dampers are fitted to provide greater passenger comfort and softer spring-type seats provide greater passenger comfort. All Series II models have track increased 1½-in. which has greatly improved the steering lock. Fully floating axles are now standard equipment.—D.A. ★

## McCAHILL TESTS

# the BRITISH LAND ROVER

*Tom calls this extremely rugged utility wagon "A great vehicle."*

### By Tom McCahill

ERNEST HEMINGWAY, crashing through the brush of Kenya in hot pursuit of a ruptured lion, would undoubtedly do so in a four-wheel-drive Land Rover. For you neophytes who go in for your Afro-adventures by listening to Three-Lips Herman and his Bongos on the local jukebox, let me assure you that the Land Rover is an established character. Gregory Peck pulled a real *faux pas* in "The Macomber Affair" when, as a celluloid White

DOUGHTY Land Rover ploughs up a storm while negotiating rocky stream bed in Bucks County, Pa.

SQUARE-BUILT rig has an expensive aluminum alloy body. A 52-hp diesel engine is optional.

15

**PLUCKY** British buggy nearly flipped over backwards this time—but luckily it didn't.

**COOL COWL** for hot weather opens up under windshield, scoops air into cabin.

**TROPIC ROOF** on Land Rover at right makes a breezeway for additional cabin cooling.

### TEST CAR SPECS

MODEL TESTED: **British Land Rover four-wheel-drive station wagon**
ENGINE: **4** cyls; **139.5** cubic ins; **77** brake hp; **124** ft-lbs max torque; **7** to **1** compression ratio. Bore **3.56** ins; stroke **3.5** ins; Fuel required: **regular.** Standard axle ratio: **4.7.** Wheelbase **88** ins; length **142⅜** ins; height **77½** ins; width **64** ins; front tread **51½** ins; rear tread **51½** ins. Weight **2,871** lbs. Gas tank capacity **10** gals. Turning circle diameter **38** ft. Tire size **6.00 X 16.**
PRICE (without optionals): **$3,160**
PERFORMANCE: 0-30 mph, **6.9** secs; 0-50 mph, **18.3** secs. Top speed **70** mph. All times recorded on corrected speedometer.

Hunter, he used an American four-wheel-drive combat vehicle as the hunt's car—simply isn't done, old boy. Such plebeian trucks are used on the garbage detail and similar tasks but never for transporting the gunners in a grade one safari. That's a job for a British Land Rover.

**The Land Rover** is to the English what the Jeep is to us. It is basically a military vehicle, adapted during peacetime to dozens of chores impossible to accomplish with a standard automobile. There are attachments available to do every job from paint spraying to snow plowing but it doesn't float too well, a feature we'll go into in a minute.

The Land Rover comes in a number of body styles and two wheelbases. It offers two sizes of station wagons and many varieties of our standard pick-up truck with canvas or solid sides and roof. Incidentally, though we didn't run a test on it, the station wagon we had featured what is known as a "tropic roof." This is a second aluminum skin fastened about an inch and a half over the regular roof. This allows a free flow of shaded air to slide over the regular roof while you are under way; inside vents scoop this breezeway-treated air right into the cab. You'd have to try it to believe it but the effect is quite cooling. Though I told the company man when he seemed to become a bit overenthused, "Good as it is, it'll never replace air-conditioning" I will concede it's the next best thing. [*Continued on page* **154**]

To test the Land Rover, a vehicle I've been interested in for years, I contacted Bill Haworth, who is doing the public relations pitch for them in this country. I've been doing stories with Bill for over a dozen years, since the days when he was public relations chief for Nash. He arrived at my house in Pennsylvania with the two Land Rovers, both 88" wheelbase. The one I was particularly interested in was the canvas-topped job which came closest to our standard American Jeep. This short-wheelbase model hasn't nearly the room of the longer (109") Rover but I wanted to compare it with my own Jeep which, though several years old now, is mechanically the same as the one Willys offers today.

**The first thing** that hits you is the price difference. The Land Rover sells for roughly $700 or $800 more than the standard Jeep. The more deluxe versions go as high as $3,600 with the long chassis. As an optional piece of equipment, any Land Rover can be had with a 52-hp diesel engine for $400 or $500 more than the standard 2¼-liter, four-cylinder petrol engine which develops 77 bhp.

The next thing I noticed when comparing the American Jeep with the Land Rover was that the Land Rover has a great deal more usable width, not only in carrying capacity but in tail room, too. Also, it has a much more costly body, made entirely of aluminum alloy. The Land Rover is sparse by any comparison but the American Jeep is far sparser. The Rover has a lot of little niceties such as a cowl which opens up under the windshield,

scooping fresh air right into your face. It even has what amounts to a spray rail behind the bumper, to toss solid water *over* the hood instead of through it when fording streams, as I did in my test. Once in the car and under way, two things become apparent immediately when you stack it against the American Jeep. First, the steering is far superior to the Jeep's not only in gearing but for ease of handling. Second, the foot room in the Land Rover must have been designed for a stunted pigmy—any guy much bigger than a barnyard rooster would resemble a pretzel after driving one of these cars on a coast-to-coast run.

**Unfortunately,** the seat cannot be pushed back, as it is unadjustable. A steel bulkhead partitioning off the rear bin locks you in there like a walnut in a nutcracker. Kelvin Fowler, the English company representative with Haworth, assured me that the 109" wheelbase job does have adjustable seats and this is something a guy my size would have to have. The rear bin, however, was ample—four hunting dog size as against the Jeep's two.

The reason we haven't any pictures of the Land Rover with the top down is because careful examination showed that to remove the canvas top and frame would be just about as simple a job as breaking out of Alcatraz with a dime-store can opener. Another fault of this car is that it needs a step to get in and out of it, even if you are a six-footer. Any gal who attempts it, unless she's wearing slacks, better know the guy she's riding with pretty well. A goof the Land Rover shares with the Jeep is the absence of an ashtray. Ashtrays are perhaps needed more in cars of this type than in any other because off the road, going through dry timber country, the lack of an ashtray might make an involuntary firebug out of almost anyone.

**The rear-view mirror** is one of those typically lousy English reducing glasses, calculated to make a big Freuhauf trailer right behind you look like a smallish beetle. The car has a good heater but it is located over the driver's shin in such a way as to keep him continually barked, unless he's "frightfully careful," as they might say. Unlike the Jeep, which uses side curtains, this number has sliding glass windows which aren't exactly the answer to a foot-weary maiden's prayer, either.

**Despite its faults,** this is still one of the world's truly remarkable pieces of transportation. In all fairness to the boys in Toledo, there's very little it can do that the Jeep can't do. The seats are slightly softer but, then, the standard Jeep seat must have been designed by one of the great-grandsons of the Spanish Inquisition torture specialists. I remember when I bought my Jeep I was ready to sell it after less than 100 miles until we located a couple of English leather bucket seats, which are in it today. Anyhow, back to the test.

Our tests were made over rugged hills, mountain trails and streams in Bucks County, Pennsylvania. Our personal safari consisted of Kelvin Fowler, the Rover factory man, and McCahill in the Land Rover, and Jim McMichael and Bill Haworth following behind in my Jeep with various lengths of chain for yanking the Rover out if we should find a spot tough enough to hang it up. Good old Bill told me it would

16

be "impossible to hang the Rover up in a place as mild as Bucks County." We have put this statement away in our "You can't do this to me" file.

At the end of a long, wooded trail there is a bulldozer-made tank trap constructed by some selfish landowner who apparently didn't want anyone to use the trail. This guy wasn't thinking of such cars as the Land Rover or Jeep but undoubtedly of passenger cars. Nevertheless, a few minutes after we started over the trap the Rover was resting on its belly with all four wheels flailing and nothing happening. It was stuck but good. Despite the embarrassment of the situation, the Jeep was brought up, chains were attached and the Rover was yanked back into the lap of civilization once again. Having established that this was not an un-stumpable vehicle, the tank trap was then approached from several different angles (meaning attacked on the bias) and was finally conquered. After this, we headed for the stream.

For a long time we traveled up the river bed with steep cliffs on each side, trying to dodge deep water holes. This terrain was as primitive as you can find in the eastern states. At one spot where we stopped, huge schools of six to 12-inch-long trout eyed us curiously and without fear from no more than six feet away in a spot where probably no line has ever been dropped—certainly not from a car. This river-cut gorge, I feel safe in stating, had never before been invaded by any vehicle, including a horse and buggy. All this was within 70 miles of New York.

At one spot, when trying to square off to climb a bank, my left rear wheel hit an almost bottomless hole. In less time than it takes a pauper to count his money, I had several inches of water over the floorboard of the cab—which is why I made the earlier statement that a Land Rover won't float for sour apples. For a moment it looked as if the only course was to desert ship. But the engine was still going, so I shot the throttle down and the front wheels (it was in four-wheel drive) pulled us out and through.

Once on dry land again we took her along trails leading through a deep woods and finally ended up in my own fields where I have a dry, steep-sided gulch only about 12 feet wide. We tried slamming the Rover down one bank and up the other. On one of these quick passages, with Haworth at the wheel (after some egging-on by McMichael and myself), it looked for a moment as if the Land Rover was going to flip over backward, but it didn't.

In summing up, the Land Rover is a great vehicle that could be made even greater with the help of a Comfort Engineer. It'll do anything a Jeep will do. It is better-built and has more carrying capacity and general room. I don't think it will do anything the Jeep won't do—at least I couldn't find anything. The test car had a top speed of about 70 mph, although any speed above 55 mph, even on turnpike-type roads, raises the discomfort index considerably and conversation must be carried on at full shout. At 50 mph it is docile and comfortable.

The Land Rover is a class vehicle from one end to the other, made by one of the most highly-respected companies in the entire industry. The regular Rover car we tested some years ago has been known to real automobile connoisseurs for years as "The Poor Man's Rolls-Royce." By the same token, this Land Rover might be classified as "A Rich Man's Jeep." I'd personally like to own the big station wagon with the adjustable driver's seat for field trial work and hunting, as the dogs would be a lot more comfortable and I could carry a lot more of them to training sessions. In a few words, this car is capable, gutty and as rugged as a cement casket.

---

**Land Rover Owner — the only magazine for Land Rover & Range Rover enthusiasts.**

Every month at newsagents, £1.95

or, available by subscription from:
LRO PUBLICATIONS LTD.,
THE HOLLIES, BOTESDALE, DISS,
NORFOLK IP22 1BZ
at £23.40 for 12 Issues.

Take an annual subscription and we will send you FREE either: a L.R.O. T-shirt or a magazine binder.
- Twelve Issues of Land Rover Owner for just £23.40 — post free and avoiding possible price increases.
- Free membership of LROC (Land Rover Owner Club) - personal membership card means many extra discounts.
- Send a cheque today and get the next 12 issues of your favourite magazine delivered to your door.

```
Please send me the next twelve issues of Land Rover Owner, starting with
the ......... issue and enrol me as an LROC member.
    I enclose cheque/PO                                     Post your complete form to:
    Charge my Visa/Access a/c no. ............             LRO Publications Ltd.,
Send me a FREE ................. (enter choice of gift)    The Hollies, Botesdale,
                                                           Diss, Norfolk IP22 1BZ.
Name ....................................... Address ...................
........................................................................
Signed ....................................................... Date .........
* Overseas rates available on request
```

**EXTRA 10% OFF FOR ARC MEMBERS — SEND ONLY £21.00**

# LAND ROVER

*British brains have put together a brawny but well-behaved passenger-utility wagon that can go anywhere you want to take it*

# WORKHORSE FOR THE CARRIAGE TRADE

### By JIM WHIPPLE

THE LAND ROVER, an incredibly sturdy passenger-utility vehicle of forest green aluminum, has become the kingpin of countless African safaris. It has replaced the pith helmet as an image of jungle exploration. Rovers ply the fabled Cairo to Capetown adventure trek as regularly and probably more easily than buses roll up and down New York's Fifth Avenue.

Now this ubiquitous, four-wheel-drive dreadnaught has turned its rugged galvanized steel bumper from beating down jungle vines to breasting the U.S. automobile market.

A practical man's first reaction to a vehicle which seems primarily designed for mechanized armies or elephant hunts might well range from cool curiosity to outright boredom. However, we predict that even a short acquaintance will leave a lasting impression and create either a desire for ownership or a wistful regret that a Rover can't fit into a particular man's way of life.

And we chose the male sex deliberately, because the LR is definitely a masculine vehicle. Muscular underpinnings replace marshmallow-soft springs, galvanized steel reinforcement substitutes for chrome trim, and simple, flat aluminum panels supplant elaborately sculptured sheet steel. This is undoubtedly the vehicle that the Marlboro man drives when he's not striking matches for the Marlboro girl.

We approached the Land Rover with the preconceived notion that it would probably turn out to be nothing more than an awkward, musclebound curiosity

*Aluminum Land Rover, with fully enclosed body, has optional tropical roof. Rugged utility vehicle has an 88-inch wheelbase, an overall length of 142 inches. It seats three in front and four in the rear compartment. All reinforcing members and fittings, like locks, steps, hinges and corner posts, are made of heavily galvanized steel, for long wear.*

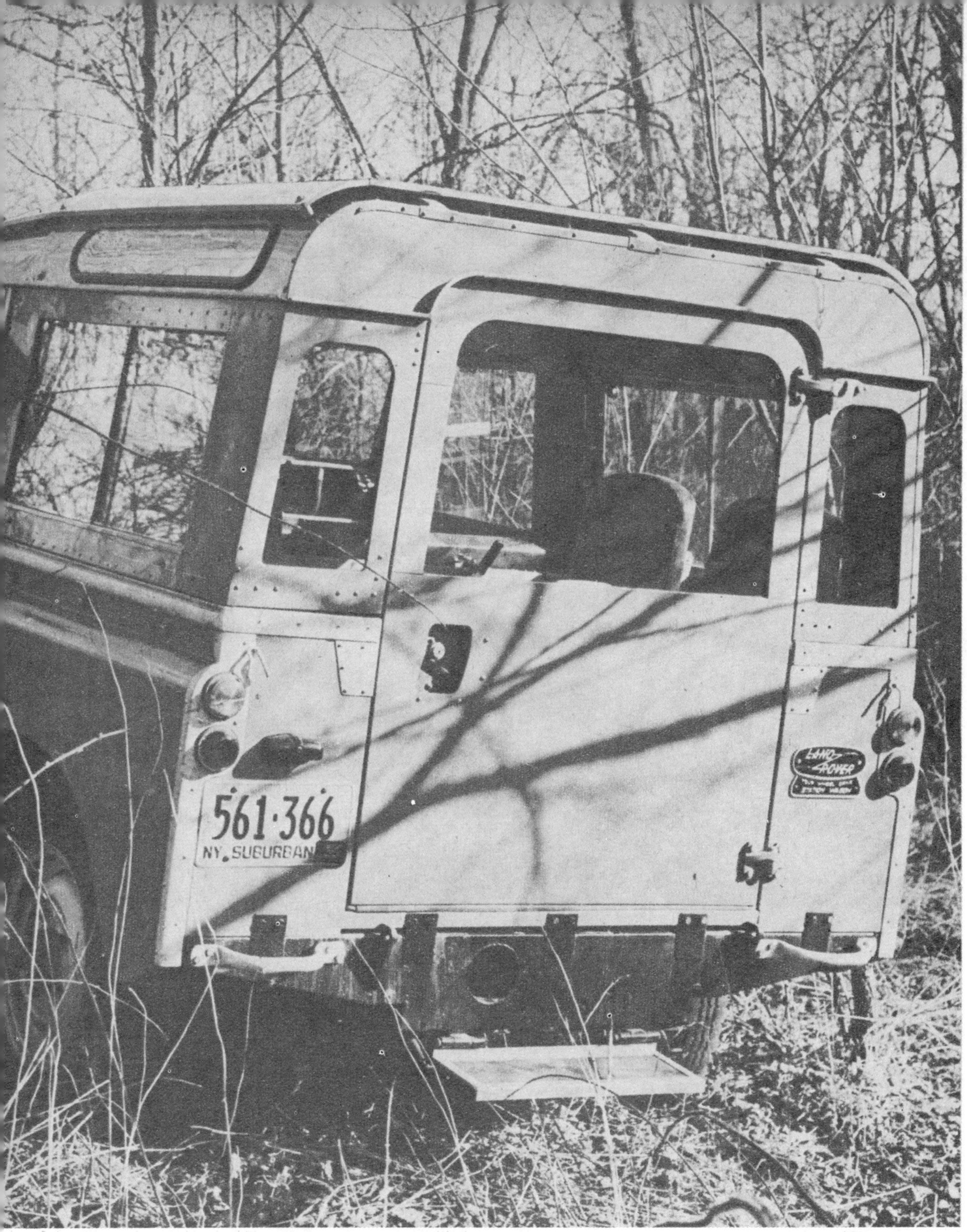
561·366
NY SUBURBAN

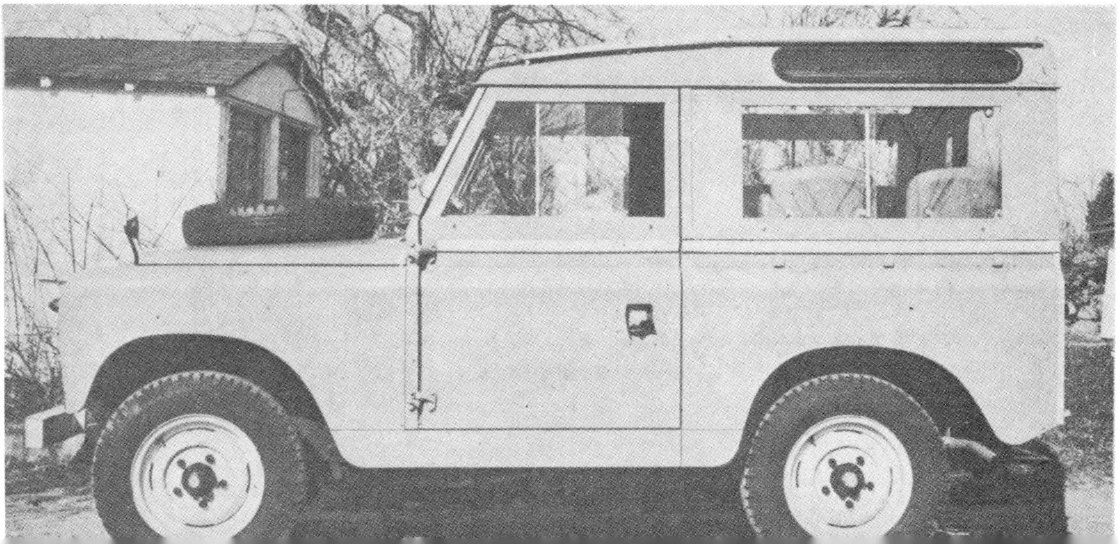

Land Rover is distinctive, easily recognized by its galvanized steel bumper, square fenders and recessed grille.

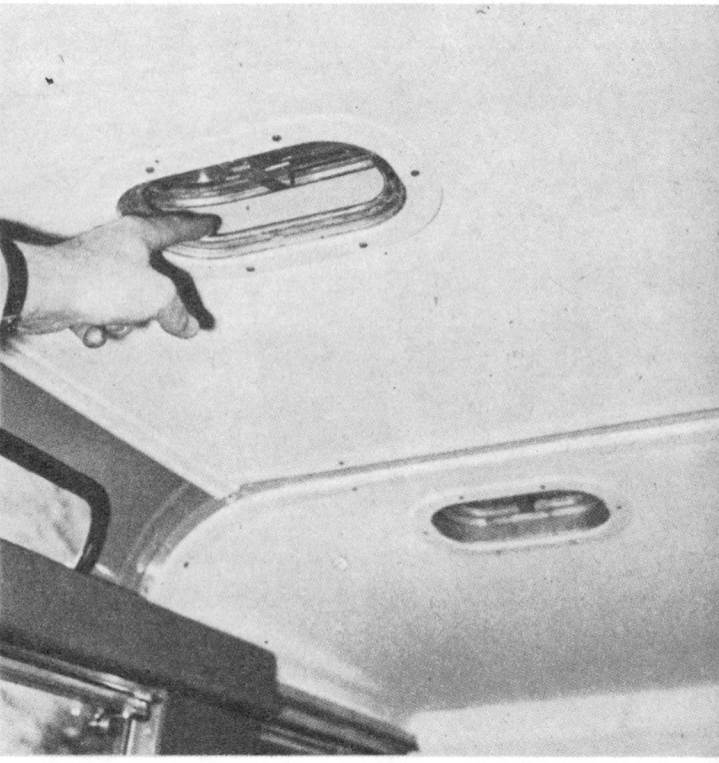

Ventilators under optional tropical roof scoop in air between tropical roof and regular roof, cooling passengers.

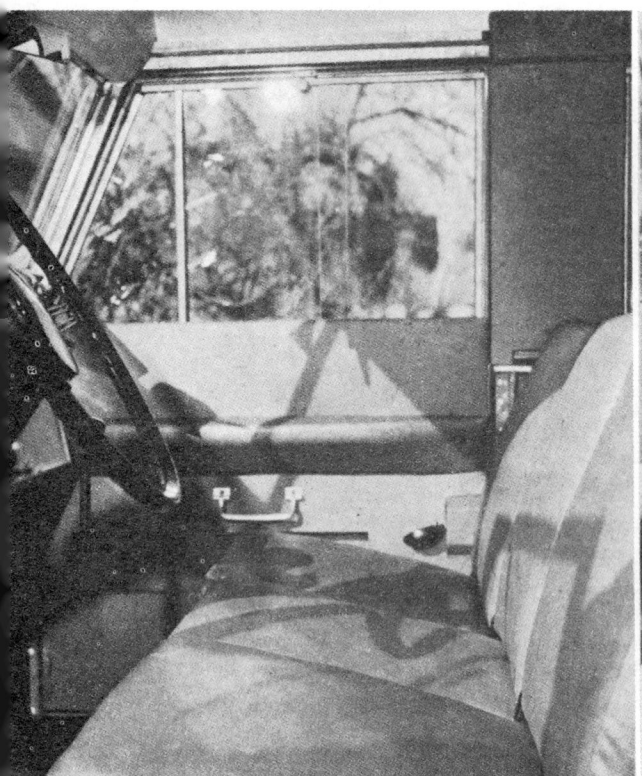

Front seat, with three separate cushions, is nearly 60 inches wide. Headroom is ample. Well grouped instrument panel includes warning lamps, electric windshield wipers. Horn, however, must be reached through spokes.

Large air vents open just above and behind hood.

Hinged rear seats are arranged at sides of LR.

Engine compartment is compact but well arranged.

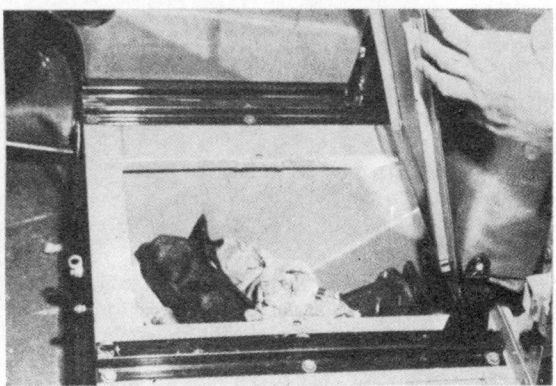

Under each seat is a separate tool compartment.

—more at home on a rhino hunt with Stewart Granger at the wheel than rolling down a country lane with the wife and kids aboard.

As things turned out our first impression proved to be way off base. Although it's true that the Land Rover is definitely the White Hunter's best friend and that this vehicle is not everyone's dish of tea—at $3,160 it's a fairly rich brew—it is also one of the most versatile vehicles to hit the U.S. countryside since the Model T Ford, and one that's a whole lot more civilized as well.

Our test LR had a full enclosed body (they call it a wagon), windows, heater, door locks, hood-mounted spare, directional signals and tropical roof. The Model 88 is extremely compact, having an 88-inch wheelbase and an overall length of just 142 inches. For reference, that's eight inches shorter than a Volkswagen sedan.

The front seat is a bench nearly 60 inches wide made up of three separate cushions which may be removed for access to the tool lockers beneath. There's ample room—and headroom too—for three people, although the middle passenger had better have short legs and small feet to locate comfortably around the shift levers.

The front seat-backs are supported by a reinforced panel running the full width of the compartment. This panel forms the lower part of the cab and prevents entry to the rear compartment via the front doors. The rear compartment contains four hinged seats arranged in pairs facing across the vehicle. They're surprisingly comfortable, although they demand bolt upright posture and force opposing long-legged passengers into a battle of kneecaps.

Exit and entry is via step and hinged rear door and is fairly easy, due to the ample headroom in the rear compartment. With the seats folded and buttoned in place, a cargo compartment 43 by 57 inches wide opens up above the wheel housings. Between the housings, the floor or bed of the cargo compartment is 36 inches wide. This enclosure is useful space that can be crammed to the roof with boxes, providing precautions are taken to prevent the load from sliding forward into the driver's seat.

Ventilation of the interior is very good, as one would expect in a vehicle designed to penetrate steaming jungles. Both doors and side panels have sliding windows that are easy to operate and which do not rattle.

The tropical roof—built in on the wagons—is a device that's disarmingly simple and remarkably effective. A sheet aluminum panel painted a light cream is mounted on short legs an inch or so above the conventional roof panel. It prevents radiant heat from hitting the roof below and permits a cooling stream of air to pass between it and the roof proper. There are four small ventilators in the lower roof panel that open forward and scoop in the air as it rushes between the upper and lower panels. These deflect streams of air downward on the passengers.

Just below the windshield there are two wide vent doors that may be opened by turning knobs. They permit a direct stream of air to enter the passenger compartment. They must be equipped with extra-cost screens to prevent the unwelcome entry of insects.

The instruments are well set up in the center of

CONTINUED ON PAGE 27

# All it
# Was a

by John F.

*Current Land-Rover Range D
Its Predecessors: 2¼-litre o.h.
and Good Economy, Whilst
ments ar*

*(Left) The Land-Rover seemed to like this sort of thing.*

AFTER several hours' gruelling test work across rough test surfaces, up and down slopes of up to 30° severity and over some wickedly rough country, all that a Series II Land-Rover needed was a wash. Good as the Land-Rover has been over its 13 years of life, this latest range is a vast improvement on its predecessors, with more power, better suspension, better brakes and improved fittings.

Although this latest series has been in production now for over three years, so great has the demand been that only now have *The Commercial Motor* been able to borrow one for a few days to test it. The wait has been worth while, however: in view of its price and capabilities the current Land-Rover represents real value for money for this class of vehicle and its overall usefulness is continually being enhanced by additions to the range of optional tools and equipment which can be used with it.

In addition to its exceptional cross-country abilities, the latest Land-Rover displays a lithe but docile performance on the road. It has a genuine top speed of 65 m.p.h. on the level, whilst on motorway down-grades over 70 m.p.h. can be reached with ease; carrying a load of 8¾ cwt., it returned 24.5 m.p.g. at low average speeds and 13.5 m.p.g. when driven at continuous full throttle for 22 miles on M1.

From a standing start 30 m.p.h. could be reached in 8 seconds and 50 m.p.h. in 21 seconds, whilst the braking efficiency was such that the vehicle could be brought to rest in 32.5 ft. from 30 m.p.h. After nearly 1,000 miles of on- and off-the-road work, no oil or water was added.

When the Series II Land-Rovers were introduced in April, 1958, the most noticeable difference between these and the Series I versions was the adoption of a new 2¼-litre overhead-valve petrol engine, replacing the previous 2-litre petrol unit, which had overhead inlet valves but side exhaust valves. Whilst the earlier engine produced only 52 b.h.p. (net) at 4,000 r.p.m. and 101 lb.-ft. at 1,500 r.p.m., the new unit develops 77 b.h.p at 4,250 r.p.m. and 124 lb.-ft. torque at 2,500 r.p.m.

*Deeply rutted tracks up a
for the Series II Land-Ro
me*

There are other important differences, however, particularly with regard to braking. Although the same size of brakes are used their performance has been remarkably improved, the stopping distance recorded from 30 m.p.h. being 14 ft. better than that obtained with a Series I version in 1957. The suspension damping has also been modified and, although I was warned that I could expect a harsher ride, this did not prove to be the case and over the rough country the vehicle handled appreciably better than the earlier model.

Body modifications have also been carried out. The external appearance has been changed by rounding off the side panels to give a waist-line effect, and by adding an apron between the chassis main members immediately behind the front bumper. The layout of the cab area is also far better. Seating has been improved by the use of spring-filled cushions and squabs,

*Despite the slippery timber surface, the fully laden vehicle reached the top of the F.V.R.D.E.'s 1-in-1.73 (30°) test slope.*

# eeded Wash

A.M.I.R.T.E.

Marked Improvement Over
ol Engine has Ample Power
nsion and Braking Improve-
worthy.

*(Right) All wheels well clear of the ground — but a smooth landing.*

*deep inclines held no terrors degree of front-axle move- noted.*

whilst the door window frames are now integral with the main door panels, so ensuring better draught-proofing. Adequate hot-weather ventilation is given by adjustable vent panels beneath the windscreens, additional ventilation being given by the sliding glass panels in the cab doors. Another minor but important change has been the use of pendant brake and clutch pedals. These reduce the number of holes in the floor and therefore make the cab more waterproof when the vehicle is wading.

So far as the chassis is concerned, apart from the alterations to the engine, brakes and suspension, its specification has not been changed greatly. The transmission line remains basically the same with a four-speed gearbox which, although it only has synchromesh on the third and top ratios, provides quick changes into second and first also, and therefore the absence of synchromesh is not of any consequence on rough work. The two-speed transfer box has the same ratios as before and separate control to enable front-wheel drive to be engaged when in the high ratio, the front drive being engaged automatically when low ratio is selected to avoid putting too much torque through the rear axle.

A wide selection of tyre sizes is offered, ranging from 6.00-16 up to 7.50-16, with the further option of Michelin "XY" 7.50-16. Cross-country tread or road-tread tyres are available. The test vehicle had Dunlop Roadtrak 6.50-16 equipment.

Complete with myself, passenger and test weights aboard, the Land-Rover was tested at a gross weight of 2 tons 1 cwt., which is just over the rated normal road capacity, which capacity should be reduced by 2 cwt. when engaged on cross-country working. Although the load was evenly distributed over the floor of the body, the rear-axle loading was 7 cwt. more than that of the front axle.

In case any damage should have ensued from the cross-country testing, I elected to carry out the normal road work first, although as it happened the unladen fuel consumption results had to be taken after the rough work, and as these tests included 22 miles of motorway work at an average speed of 62 m.p.h., it is fairly safe to assume that none of the chassis components suffered any damage.

The results obtained on the road are detailed in the accompanying data panel and are fairly self-explanatory. The low-speed consumption runs were made over a 7.5-mile undulating country road in each direction, whilst the motorway run also was 11 miles north and 11 miles south to minimize any errors due to wind resistance. Even so, there was a very strong cross-wind when making the unladen run, which had the effect of reducing the speed, although this lower speed would be partly because of the reduced momentum of the unladen vehicle. Nevertheless, there was comparatively little difference between the fuel

*This effect is obtained by entering a 1-ft.-deep muddy pool at over 30 m.p.h. The engine did not falter and the body remained watertight.*

## ROAD TEST No. 715/M174—LAND ROVER SERIES II ½-TON 4 × 4

**MODEL**: Land-Rover Series II Regular 88-in.-wheelbase four-wheel-drive general-purpose vehicle, with standard petrol engine.

**WEIGHTS**: 
|  | Tons | cwt. | qr. |
|---|---|---|---|
| Unladen (kerb weight) | 1 | 7 | 3 |
| Payload | | 8 | 3 |
| Driver, observer, etc. | | 4 | 2 |
| | 2 | 1 | 0 |

**DISTRIBUTION**: 
Front .. .. .. 17 0
Rear .. .. .. 1 4 0

**ENGINE**: Rover four-cylindered o.h.v. petrol engine; bore 90·49 mm. (3·562 in.); stroke 88·9 mm. (3·5 in.); piston-swept volume 2·286 litres (139·5 cu. in.); maximum net output 77 b.h.p. at 4,250 r.p.m.; R.A.C. rating 20·5 h.p.; maximum net torque 124 lb.-ft. at 2,500 r.p.m.

**TRANSMISSION**: Through 9-in.-diameter single-dry-plate clutch to four-speed synchromesh main gearbox and two-speed transfer box thence by one-piece propeller shafts to the fully floating spiral-bevel front and rear axles.

**GEAR RATIOS**: Main box: 2·996, 2·043, 1·377 and 1 to 1 forward; reverse 2·547 to 1; transfer box; 2·888 and 1·148 to 1; axle ratio 4·7 to 1.

**BRAKES**: Girling hydraulic system with leading-and-trailing-shoe units at all wheels. Single-pull handbrake linked mechanically to 9 in. × 1·75 in. drum brake on rear of transfer box. Diameter of drums, front, 10 in., rear, 10 in., width of linings, front, 1·5 in., rear, 1·5 in.; total frictional area 104·7 sq. in., that is, 53·8 sq. in. per ton gross weight as tested.

**FRAME**: Welded box section with five cross-members welded in position.

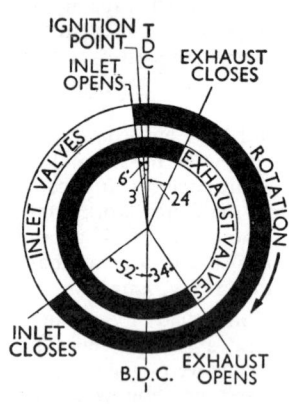

FIRING ORDER 1·3·4·2
VALVE CLEARANCES 0·010"
COMPRESSION RATIO 7:1

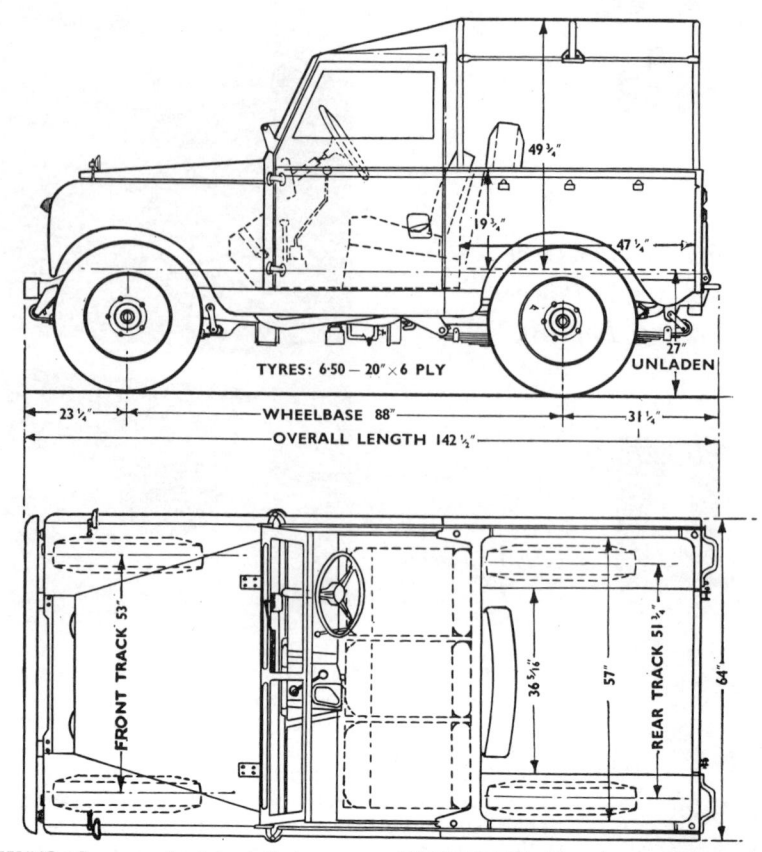

TYRES: 6·50 — 20" × 6 PLY
WHEELBASE 88"
OVERALL LENGTH 142½"
FRONT TRACK 53"
REAR TRACK 51¼"

**STEERING**: Burman recirculating ball: 4 turns from lock to lock.

**SUSPENSION**: Semi-elliptic springs, with telescopic dampers at both axles.

**ELECTRICAL**: 12 v. compensated-voltage-control system with 57-amp.-hr. battery.

**FUEL CONSUMPTION**: (a) laden, low speed, 24·5 m.p.g. at 32 m.p.h. average speed; (b) laden, high speed, 13·5 m.p.g. at 63 m.p.h. average speed; (c) unladen, low speed, 27·4 m.p.g. at 32 m.p.h. average speed; (d) unladen, high speed, 13·7 m.p.g. at 62 m.p.h. average speed, that is 47·8 gross ton-m.p.g. as tested (a) and 26·3 gross ton-m.p.g. as tested (b), giving time-load-mileage factors of 1,530 (a) and 1,657 (b).

**TANK CAPACITY**: 10 gal., normal-speed laden range approximately 240 miles.

**ACCELERATION**: Through gears, 0–20 m.p.h., 4·0 sec.; 0–30 m.p.h., 8·0 sec.; 0–40 m.p.h., 13·5 sec.; 0–50 m.p.h., 21·0 sec.; top gear, 10–20 m.p.h., 7·25 sec.; 10–30 m.p.h., 13·25 sec.; 10–40 m.p.h., 20·0 sec.; 10–50 m.p.h., 29·25 sec.

**BRAKING**: From 20 m.p.h., 14·25 ft. (30·2 ft. per sec. per sec.); from 30 m.p.h., 32·5 ft. (29·7 ft. per sec. per sec.). Handbrake from 20 m.p.h.; (a) rear-wheel drive only, 48 per cent. (Tapley meter); (b) four-wheel drive engaged, 100 per cent. (Tapley meter).

**WEIGHT RATIO**: 1·97 b.h.p. per cwt. gross weight as tested.

**FORWARD VISIBILITY**: To within 11·5 ft. of front bumper at ground level on centre line.

**TURNING CIRCLES**: 41 ft. left lock, 40·5 ft. right lock. Swept circles: 43 ft. left lock, 42·5 ft. right lock.

**MAKERS**: The Rover Co. Ltd., Solihull, Warwickshire.

---

consumption rates laden and unladen when running under continuous full-throttle conditions.

The acceleration results I obtained can be considered highly satisfactory for a vehicle of this type, and although top gear was engaged before 50 m.p.h. was reached, this speed can be obtained in third gear. The times recorded in top gear between 10 and 50 m.p.h. were entirely satisfactory also, although slight transmission roughness was noted between about 10 and 15 m.p.h. Nevertheless, the engine pulled extremely well at low speeds, which is particularly satisfactory in view of the relatively high speed at which maximum torque is developed.

In connection with speeds, it is of interest to note that the standard speedometer fitted to this Land-Rover was one of the most accurate that I have ever come across. Timed checks at up to 60 m.p.h. showed the meter to be accurate to within 1 per cent., as was the mileage recorder also.

When making the foot-brake tests from both 20 m.p.h. and 30 m.p.h., all the wheels locked evenly without any signs of uneven pulling. Two handbrake tests were carried out from 20 m.p.h., the first being with rear-wheel drive only engaged, and a very good retardation was recorded. Then I engaged front-wheel drive and on each occasion the Tapley meter registered over 100 per cent., which figure had also been obtained from each of the foot-brake testing speeds.

Engine cooling and brake-fade resistance tests were made on a ¾-mile hill which has an average gradient of 1 in 10½. At the time of the tests the ambient temperature was 62° F. and before making the climb the engine-coolant temperature was 158° F. The ascent was made in just over 1¾ minutes and at the top of the hill the water temperature was 165° F., at which temperature the thermostat had just started to open. The minimum speed during the ascent was 23 m.p.h. and the lowest gear used was third, this being engaged throughout most of the climb.

To test for fade I coasted the vehicle down the hill in neutral, engaging top gear and applying full throttle towards the bottom of the slope to compensate for the lack of gradient. This descent was made without exceeding 20 m.p.h. and it lasted 2 minutes 50 seconds. At the bottom of the hill a "crash" stop was made from 20 m.p.h. and the Tapley meter showed the maximum efficiency of the brakes to be 97 per cent.—this is only 3 per cent. lower than was recorded with the drums cold, showing that fade should not present any problems with this vehicle. Similarly, the slight rise in the water temperature during the preceding climb proves that engine cooling is entirely adequate for all conditions.

The vehicle handled extremely well on the road at all speeds and was comfortable both to drive and to ride in as a passenger. All the controls were light, the steering being very good with no signs of wander at any speed. The castor action on the vehicle tested was a bit sluggish at first, but as the mileage increased the steering freed off. Large single undulations in the road surface can create some steering-wheel shake, but generally the steering geometry is entirely satisfactory.

Because of the cross-country tyres, the Land-Rover tested was by no means quiet to ride in, but with normal road-tread tyres this should not be the case although, because all the gear combinations are indirect, there will always be some transmission whine.

Under both laden and unladen conditions the suspension behaved admirably on all types of road surface. Slight pitching can be set up, as is expected with any short-wheelbase vehicle, but roll is negligible and the Land-Rover corners more like a sports car than a cross-country 4 × 4.

Although the standard driving seat is not adjustable its position strikes a good average and the squabs and cushions are comfortable. A slight criticism concerning the driving position is that, with the driver's left hand on the left-hand upper section of the steering wheel, the speedometer dial is almost completely obscured.

**Large Fuel Filler**

Other good details about the latest Land-Rover are the well-positioned wing mirrors and the large-diameter fuel filler, which has an extending neck to permit filling from cans. A large captive cap is fitted. The optional heater and demister unit, although of the recirculatory type, is very effective, but the separate windscreen wipers are a little slow by modern standards.

A fault which may or may not have been peculiar to the test vehicle was its rather bad starting, although on no occasions were the mornings cold, and starting was even a little dubious when the engine was quite hot. When warm also, particularly after a cross-country run, the engine displayed an annoying tendency to run on after the ignition had been switched off. A little attention to either the carburation or ignition could possibly have rectified these faults however.

For the off-the-road working I was allowed to make use of the Ministry of Aviation's facilities at the Fighting Vehicles Research and Development Establishment, Chertsey.

Maximum gradient performances were assessed first on the four test slopes, the most gentle of which is 1 in 4. Facing up this slope the handbrake held the Land-Rover satisfactorily, but a restart in bottom gear, high auxiliary did not quite come off, although an easy restart was made in third gear, low auxiliary. Facing down the slope, an easy restart was made in reverse-low and once again the handbrake held the vehicle with ease.

On the next slope, where the gradient is 1 in 3, again an easy reverse-low start was made while facing down the slope, but third-low proved a little too high for the Land-Rover when facing up this hill and second-low had to be employed.

The other two test hills are both timber surfaced and because light rain was falling at the time of my tests—and had been falling for some hours previously—these surfaces were treacherously slippery. After several attempts on the 1-in-2 slope this gradient was eventually surmounted using second-low, the successful ascent including a second-low restart almost at the top after wheelspin had brought the vehicle to a halt. Because of the slipperiness of the slopes, I reduced the tyre pressures from their normal figures of 25 p.s.i. front and 30 p.s.i. rear to 20 p.s.i. all round, and this immeasurably helped to improve traction.

Both facing up and down this hill the handbrake held the vehicle safely and when facing down at the bottom of the slope a satisfactory reverse-low restart was made.

I then turned my attention to the steepest of the four hills, the gradient of which is 1 in 1.73. With bottom-low engaged we almost managed to reach the top of the hill until wheelspin brought us to a stop. The tests carried out lower down the hill showed that there was sufficient power to enable a restart to be made on this gradient and that it was only the slipperiness of the timber baulks which was

*The roughness of the F.V.R.D.E.'s pavé track can be judged from this picture.*

preventing a successful ascent. Similarly, a smooth restart was made in reverse while facing down the hill and in both conditions the handbrake was powerful enough to hold the vehicle.

It should be pointed out that although I ought to have reduced the payload by 2 cwt. to comply with the manufacturer's recommendation, all these and subsequent rough country tests were made with the full on-the-road payload at a gross weight of 2 tons 1 cwt.

The F.V.R.D.E. has three gruelling suspension courses, each of which is 300 yd. long. The Land-Rover was driven over the pavé course quite smoothly at about 40 m.p.h., and although there was plenty of wheel movement, body movement was not so great and there were very few rattles. The ride was definitely above average, although it was necessary to hang on to the steering wheel pretty grimly to maintain a straight course.

The other two suspension tracks are paved with raised setts, these setts being staggered so as to vary the periodicity. The first track has setts 1 to 1½ in. high and so well did the suspension cope with this course that I was able to drive the Land-Rover down the track at 45 m.p.h. without holding the steering wheel at all.

The other sett track has 2-2½-in. blocks and over this the ride was a bit rougher, but still very smooth, and at 40 m.p.h. again I was able to take my hands off the wheel.

When a Series I Land-Rover was tested in 1957, it was possible to maintain only 15 m.p.h. over this second sett course, and I remarked at the time that the vehicle was only just controllable and sometimes turned through 45° to the direction of travel.

Following these " hard-surface " trials, the Land-Rover was then taken to the rough-road course at Bagshot Heath and the accompanying pictures give some idea of what the 4 × 4 was put through, again with the full road payload. On numerous occasions all four wheels left the ground, but landing did not disturb the occupants of the vehicle or the vehicle itself overmuch.

During this sort of manœuvre three years ago, I managed to bend the front axle of the Series I Land-Rover, but nothing like that happened this time, the front axle having been reinforced by a welded channel-section underneath it to resist this sort of thing.

Loose-surfaced gradients of up to 1-in-2 severity were encountered at Bagshot Heath, but none proved too much for the Land-Rover, whilst general waterproofing was proved by several high-speed dashes through deep puddles.

All in all, most satisfactory performances which the manufacturers may be justifiably proud of.

To complete the picture, I did a few maintenance tasks on the vehicle using the tool kit provided by the manufacturers. Whilst this is fairly comprehensive, strange omissions are the lack of spanners to fit the rear-axle oil-filler plug and the handbrake adjusting screw. The poor-quality jack is nothing to be proud of either.

I tackled the under-bonnet jobs first, raising the bonnet taking only 2.5 seconds, a permanent hinged stay being provided to hold the bonnet in the raised position. With the bonnet up, I checked the water level in 5 seconds, the engine oil level in 8 seconds, the level of the hydraulic fluid for the clutch and brake circuits in 5 seconds, the battery levels in 29 seconds, the steering box level in 55 seconds, the air-filter oil level in 2 minutes 5 seconds and the contact-breaker points gap in 1 minute 35 seconds.

I cleaned out the sediment bowl adjacent to the petrol-lift pump in 1½ minutes and the gauze filter at the carburetter intake union in 42 seconds. It took me 50 seconds to change a fuse because the fusebox is a little awkward to reach. Number 4 sparking plug was removed in 1 minute 40 seconds and replaced in 42 seconds, the time of removal being extended because the plug was tight to turn. Following all this, the bonnet took me only 2.5 seconds to close and secure again.

Working from underneath the vehicle, I checked the oil levels in the main and transfer gearboxes in 2¾ minutes and the level of the oil in the front-axle differential in 45 seconds, a further 1½ minutes being required to check the level in each of the two swivel-pin housings. Because there was no suitable spanner, I could not check the level of the oil in the rear axle.

Working with the crude mechanical jack supplied in the tool kit, I then adjusted the four wheel brakes, these taking a total time of 9½ minutes jacking each wheel up separately. There is only one adjuster per brake and this is easy to reach. The separate transmission handbrake was adjusted in 1 minute 15 seconds, using an adjustable spanner.

As a final task, I removed the spare wheel from its stowage behind the front seats in 23 seconds and replaced it in 50 seconds. It is clamped in place by a bar secured by a sensibly sized wing nut.

The standard Series II Land-Rover as tested retails at £660, which makes good value for money in anybody's language. There is also a diesel-engined version, the current price of which is £760, but there should be some news of another development in this direction in the course of the next few weeks.

# There's a welcome for all off-roaders in the AWDC

- Wide variety of vehicle interest
- Largest UK calendar of off-road motorsport
- Quality All Wheel Driver magazine
- Club Shop
- And, best of all, a friendly spirit.

BRITAIN'S CROSS-COUNTRY VEHICLE ASSOCIATION
*Established 1968*

**AWDC**
ALL WHEEL DRIVE CLUB

FULL DETAILS (SAE please) FROM — AWDC, P.O. BOX 6, FLEET, HANTS, GU13 9YY

CONTINUED FROM PAGE 21

the panel and have efficient black-on-white dials with good edge lighting.

On the plus side is an unusual warning lamp designed to light up when engine has reached normal operating temperature if the mixture control (choke) knob remains open in the starting or warmup position.

Transmission control is by central lever with the pattern of the four forward speeds embossed on it. Front wheel drive clutch lever and low range shift lever are color coded (red and yellow). Starter button is located below the instrument panel and to the right—convenient for right-hand-drive English users but a long reach from the left side. The horn button is an extension from the steering column below the wheel and must be reached through the spokes. There are two powerful electric windshield wipers that really do a job, as well as an efficient (and noisy) recirculating heater with excellent demisting and defrosting action.

In spite of its rugged appearance, the Land Rover 88 is easy to handle in traffic and on the highway. Thanks to recirculating ball gear, the steering is light and fairly precise, except when the vehicle is at a near standstill.

Brake pedal pressure is light, yet brake action is powerful even at higher speeds. The clutch pedal is coupled hydraulically to the throwout lever and has a crisp, direct action, although the heavy-duty clutch requires more than average effort to disengage it.

Third and top gears are synchromesh so that shifts from second to third and third to top (fourth) and back to third can be made smoothly. Downshifting to second and first requires double clutching.

Two features of the Land Rover's performance make an immediate and favorable impression. First is its amazing roadability and handling. In spite of its high structure and short wheelbase, the LR takes sharp curves smoothly with no sway, very little lean and a definite feeling of security. All four wheels stay exactly where you want them to at all times.

One clue to the LR's stability may be the aluminum construction, which keeps the upper part of the vehicle (body panels, fenders, roof, hood, etc.) light, and concentrates the weight down around the engine, frame and running gear.

Second are the car's riding qualities —remarkably good for a road and rough-terrain vehicle. The LR is not noticeably less comfortable than the average automobile on highways and expressways as far as front seat passengers are concerned. On bumpy, narrow, blacktop roads the well-controlled suspension gives a firm ride— but also provides a great deal of freedom from bouncing and pitching. The LR will cruise easily at 65, although the greater noise level—in comparison with standard passenger cars— makes 50 mph cruising much more relaxing.

Driving position in the LR is good, and the steering wheel is properly located. Unfortunately, the "88" model lacks an adjustable driver's seat and foot room around the control pedals. The accelerator pedal offers no way for the driver to rest or brace his foot comfortably, which means considerable fatigue on any trip longer than a few miles. It seems a shame that such relatively minor details— from an engineering standpoint— should mar the pleasure of long-distance driving in an otherwise very satisfactory vehicle.

In off-the-road work the LR is a real champion. Its rugged frame would do credit to a bridge builder, and its lowest gear multiplies the power of its 77 horsepower engine by a 40-to-1 ratio. Springs and axles are very heavy for the size and weight of the vehicle, and in four-wheel drive and low gear range it is next to unstoppable.

The floor and undercarriage are high and the few items projecting below the frame level are extremely sturdy. A snapped-off sapling couldn't harm the muffler and tailpipe, which are rugged enough to serve an army tank.

The body is of aluminum with all reinforcing members and fittings, such as locks, steps, hinges, corner posts, etc., made of steel which has been heavily galvanized to prevent corrosion. Frame and running gear are protected by heavy black paint. The LR is clearly a vehicle that could be garaged outdoors for its lifetime with little or no ill effects.

What is the LR's place in the American automotive picture? Well, we think that although somewhat specialized, it's a long way from being limited in usefulness to cattle ranches and Rocky Mountain goat hunts. As a vehicle for a single-car family, however, it could be recommended only to a few farm families with a very moderate amount of long-distance driving to do.

However, as a second car for suburban and country families with a lot of short-trip operation or hunting, camping and fishing expeditions, plus a need for either a station car to get through no matter what, or a light pickup truck, the LR would really pay its way. Its usefulness on the farm could easily be increased by hooking on a two, or even a four-wheeled trailer, which the LR has ample traction and power to handle in either on or off the highway hauling.

Although initial cost is fairly high, there is absolutely no style obsolescence problem and the Land Rover is engineered and built to last for a long, long time. Properly maintained, with attention to essentials of lubrication, brakes, clutch, tires and engine tuneup, a Rover should be just as serviceable at the end of its tenth year as at the end of its first.

Such a promise of usefulness and extended ownership brings us to the conclusion that it will probably cost you no more to operate this first-class safari vehicle down on the farm than it would be to buy a used sedan or pickup every two or three years. •

# LAND ROVER Station Wagon

► "Don't worry, we'll make it!" shouted the driver each time I clawed wildly to brace for the rollover that seemed imminent. We were heading up a precipitous trail, crawling slowly up the crest of a razorback peak with just enough firm ground underneath to support the four wheels.

At last we reached the top and a miniature plateau. I forgot for a moment what lay behind and took in the views to the left, a range of higher mountains, while far to the right the Pacific shone dully through a shroud of haze. It seemed incredible that minutes before I had been sitting comfortably in the Pasadena showroom of Peter Satori, Inc., listening to an explanation of the Land Rover station wagon. Then when Rover Division manager Joe Krumbolz volunteered a brief demonstration ride, I accepted. And here we were.

Nothing for that now. "How do we get out of here?" I asked Joe, now obviously in his element. His reply was to

How's this for a functional cockpit layout? Lockable storage cabinets are under each chair-height seat. There are three central shift levers!

engage first gear-low range and aim carefully for the apparent edge of a sheer cliff.

Down we plunged, making haste slowly. Without any official measurement, just say the grade was steep; it appeared to be a wild 45 degrees. Yet with the low range first gear's 40-to-one overall ratio yielding a mere 2.2 mph per 1000 rpm, it was unnecessary to touch the brakes. Only when we nosed over into a still steeper downturn was Joe forced to stab quickly at the brake pedal to slow us.

Very quickly we arrived at the bottom in safety with a sigh of relief. Thus ended my first ride in a Land Rover. After reflection, I realized that this was no gimmick thrill machine; it was and is a muscular vehicle that will take just about anything man can deliver. And after a few hours behind the wheel, I became convinced that the only limit to the vehicle is the bravery (or foolhardiness) of the driver. Better yet, it's a tremendous kick to drive.

What makes it so? As in most truly efficient vehicles, it is a combination of many things. The Rover Company started from scratch (there is no relationship to their passenger models) to design and build what they call "the world's most versatile vehicle." Versatile in this context means a machine that is suited for normal highway use, but finds its true *métier* plowing across desert sand or streaming jungle, poking into normally inaccessible spots anywhere on the globe.

Versatile also means an amazing choice of vehicles from two wheelbases (88 and 109 inches), and either a 51-bhp, 123-cubic-inch Diesel or a 77-horse, 139.5-inch gasoline engine. We tested the most elaborate civilian model they make, the long-wheelbase station wagon, which is available only with the gasoline engine. The list of optional body configurations includes a basic pickup with a canvas top, the same with a hard-top cab, plus a full length hardtop with or without side windows, and two each station wagons and fire engines.

It's trite to say that the Land Rover is rugged. What else could it be for its purpose? The chassis frame and crossmembers are built up from flat plates into very deep box sections. With a 9¾-inch ground clearance it is nearly impossible to hang the Rover in a rut. Though that much open space below coupled with an 81-inch overall height indicates a top-heavy machine, it's not so. The body panels are constructed of light alloy so that weight above the frame is held to a minimum. Steep side hills are thereby safe to traverse.

The alloy body is one of the machine's fascinations. Straightforward in the extreme, it is a mass of exposed rivets and hinges, galvanized steel reinforcing plates, handles, knobs and sliding windows. The little functional touches tell of the thought that entered its design. For example, tiny bits of plastic tubing inserted in the body just below the six sliding side windows carry rain water from the window channels to the outside; all the door handles are recessed into the body so that one may travel through thick underbrush without getting entangled. In fact, there are no projections on, under or above the body. Grille and headlights are deeply set back from the fenders for pushing through brush country and the front bumper is a ⅛-inch-thick galvanized steel box section that appears capable of bowling over small trees. Exposed members underneath are also extra-heavy-duty.

Short of air conditioning, the best substitute is the unusual tropical top. It is an additional alloy panel mounted with about an inch of air space between it and main roof. This reflects much of the sun's radiant heat and allows the four vents in the roof itself to be opened for fresh air without allowing rain inside the vehicle.

Everything is related to its function. There was no waste motion in an attempt to design an elaborate interior. For example, a pair of windshield wiper motors stare right back at one from the bottom center of each half of the split glass. Need a little more fresh air than is provided by the sliding windows and the roof vents? Just turn the two large black knobs above the dash. Each controls a rectangular trap door that opens out just below the windshield. Screens are optional to keep out the tsetse flies. At the base of the dash is a convenient hand throttle on a notched quadrant. Conceivably, the going could get so rough that one couldn't hold the right foot steady enough.

Upholstery is a serviceable gray plastic on the bench seats. The headliner, in startling white plastic, contrasts sharply. A twin dome-light system reflects off it, lighting the interior most satisfactorily. At night you can read a map, comfortably, in any seat in the wagon. Under the front seats are two covered lockers that can be secured with padlocks; under the back seats there is substantial storage for tools and incidentals.

Three-abreast front seating is possible but the center man must look sharp to avoid the trio of gear control levers. Their scheme is so: the big lever on the center of the tunnel engages a conventional four-speed box; to its right is a stubby lever with a bright yellow knob; a few inches more right is another short lever, this one topped in red. Push down on the yellow and one has engaged four-wheel drive —

this may be done while moving. Pull back on red and low range takes over, four-wheel drive automatically engaging. The vehicle must be stopped to select low range. From our gear ratio chart the variety is evident. Part of the fun of driving is charging up impossible slopes and then experimenting to find the best of the twelve forward combinations for the job.

Many American owners who do significant mileage on highways in rear wheel drive find it worthwhile to install Warn hubs at the front to improve the free-wheeling. (Warn Mfg. Co., Riverton Box 6064, Seattle 88, Wash.)

Only 77 bhp for 3740 pounds means that acceleration isn't going to be much — and it isn't. It really isn't too important that the Rover only hits 49 mph in the standing quarter-mile unless you're trying to out-race charging elephants. A comfortable highway tempo is 65 mph; it can be maintained for hours without strain. Fuel economy is not sensational; we realized a range of 11-14 mpg on Mobilgas regular during our rugged test.

Vitals under the hood are easy to reach. The four-cylinder o.h.v. engine is fractionally oversquare, developing its 124 lb-ft of torque at only 2500 rpm. Horsepower peaks at a mild 4250 revs and from the sturdy way in which the engine is laid out it should prove trouble-free for a long period with only nominal care.

Steering is worm and nut with recirculating ball. It is easy to manipulate while moving, but it's a man-sized job to parallel-park the Land Rover. We understand the necessity for four turns of the wheel from lock to lock.

After a look at the four beefy semi-elliptic springs that are the heart of the non-independent suspension, it is safe to assume that the resultant ride is firm. It is comfortable along the highway but be assured that there is no wallowing through dips.

This is the only station wagon that we have ever seen that will really carry 10 adults. Front and center seats carry the usual three each. The rear seats — enter through the back door, please — fold down from each side and allow four people to face each other, working out among themselves where all the knees go. Fold the center seat forward, lash the rear seats in their raised position and presto: a long cargo area just like American wagons.

Three persons can pack in on a long safari with a ton of supplies in this model. If it's a luxury safari they might consider utilizing the three power-takeoff points available. A winch may be fitted at the front (a must for swamp country) while inside, the center takeoff will drive generators for lighting, radios, air conditioning or whatever is desired. The rear power point can be fitted with pulley drive for saws, mowers, threshers, etc.

It's not a cheap plaything, the Land Rover wagon. From a West Coast base price of $3880, the added cost of heater, water and oil gauge, hand throttle and second windshield wiper brings it to $4095 plus tax and license. In relation to what's comparable, this is costly. The Japanese Toyota Land Cruiser, 135 bhp, lists for $3365 but it has no transmission low range; DKW makes a much smaller Jeep-type four-wheel-drive machine for $2995; the Willys Jeep, refined through many years, is based at about $2500 for a 101-inch-wheelbase model but it won't carry as much or as many as the Land Rover. The Rover boys build a lot of quality into their package. You won't see many on Park Avenue or Wilshire Boulevard, but in pith helmet or parka country a Land Rover is the height of fashion. —*Wayne Thoms*

---

**LAND ROVER Station Wagon**

Price as tested: $4095 (West Coast)

Importer: Rover Motor Co. of North America Ltd.
36-12 37th Street
Long Island City 1, N. Y.

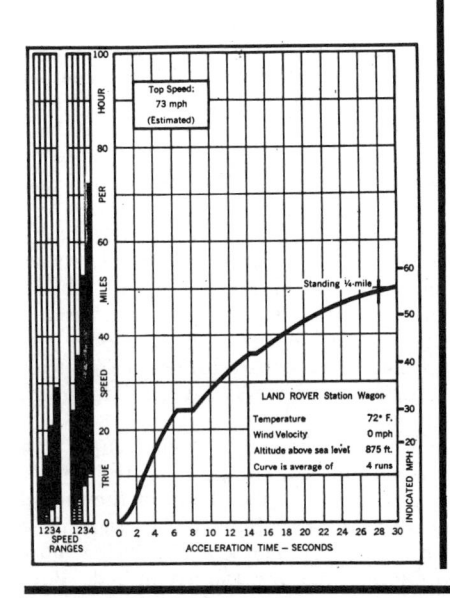

**ENGINE:**

Displacement.................139.5 cu in, 2286 cc
Dimensions..Four cyl, 3.56 in bore, 3.50 in stroke
Valve gear..Pushrod overhead valves, roller tappets
Compression ratio...................7.0 to one
Power....................77 bhp @ 4250 rpm
Torque..................124 lb-ft @ 2500 rpm
Usable range of engine speeds.....650-4500 rpm
Corrected piston speed @ 4250 rpm....2500 fpm
Fuel recommended....................Regular
Mileage........................11-14 mpg
Range on 19 gallon tank..........210-270 miles

**CHASSIS:**

Wheelbase ..........................109.0 in
Tread ..............................51.5 in
Length .............................175.0 in
Ground clearance......................9.8 in
Suspension: F & R, rigid axles, longitudinal semi-elliptic underslung leaf springs.
Turns, lock to lock....................4.0
Turning circle diameter between curbs......45 ft
Tire and rim size............7.50 x 16, 16 x 6
Pressures recommended: 22 psi normal, 22/36 psi heavily loaded
Brakes; type, swept area..11 in drums, 311 sq in
Curb weight (full tank).................3400 lbs
Percentage on driving wheels.............100%
Test weight ........................3740 lbs

**DRIVE TRAIN:** (Includes two-speed transfer box)

| Gear | Synchro? | Ratio High-Low | Step | Overall High-Low | Mph per 1000 rpm |
|------|----------|----------------|------|------------------|------------------|
| Rev | No | 2.92-7.35 | | 13.75-34.59 | 6.4-2.5 |
| 1st | No | 3.44-8.66 | 47% | 16.17-40.69 | 5.4-2.2 |
| 2nd | No | 2.34-5.90 | 48% | 11.03-27.74 | 8.0-3.2 |
| 3rd | Yes | 1.58-3.98 | 37% | 7.44-18.71 | 11.8-4.7 |
| 4th | Yes | 1.15-2.89 | | 5.40-13.58 | 16.1-6.4 |

Final drive ratio: 4.70 to one, front and rear.

Top Speed: 73 mph (Estimated)

Temperature 72° F.
Wind Velocity 0 mph
Altitude above sea level 875 ft
Curve is average of 4 runs

# LAND-ROVER 1961 PETROL MODELS

## "MOTOR TRADER" Service Data

THESE vehicles have been in production for a number of years, and have been the subject of previous articles in this series. Design developments have taken place in a progressive fashion and so it will be apparent that while, outwardly, the vehicles have changed but slightly during their 12-year production period, inwardly and mechanically there has been considerable change.

Present series vehicles were first marketed in 1958. They were offered in either 88in or 109in length wheelbase versions, and with the option of either 2¼-litre petrol engines or 2-litre diesel engines. Our last article on Land-Rover vehicles featured the diesel-powered version and to complete servicing information available to readers, we are presenting this data sheet which not only deals with the petrol-engined vehicle but details changes of service procedure which are common to both versions.

Apart from the engine options, much of the original basic mechanical layout of the earlier series is retained, with the exception of the front wheel drive shafts and some component parts of the front suspension; Hardy Spicer joints replace the Tracta type universal joints and king pins have a revised mounting. Gearbox and transfer box are much the same as the earlier series, and differential units are closely comparable with those used in current production Rover cars.

Vehicles are identified by nine figure serials. First three digits represent vehicle engine and specification type, fourth digit the year or sanction period, i.e., 8=1958, and the last five digits denote actual serial number of the vehicle. This serial number will be found stamped on the transfer box instruction plate on the dash panel over the gearbox cover and is the same as the chassis number stamped on the right-hand front spring shackle bracket. Engine serials are to be found on a boss on the left-hand side of the cylinder block at the front. The vehicle serial number should always be quoted when referring to the makers or when ordering spare parts.

The few special tools required for service are available from the Rover Co., Ltd., and are listed in these pages.

Threads and hexagons are in the main of the Unified series.

**DISTINGUISHING FEATURES.** Similar in general appearance to its predecessors, this model has side lamps recessed into front wings and flush-fitting exterior door handles

## ENGINE

### Mounting

At front, angle brackets are bolted up to bosses either side of engine unit and to chassis frame by two ½in UNF bolts and lockplates each. At rear, gearbox transfer casing unit rests on brackets attached to either side of casing by four studs and nuts each and to frame by retaining bolt and nut together with adjuster plate and locknuts. Tighten all bolts fully.

### Removal

Remove engine without gearbox. Disconnect and take off bonnet, drain coolant, remove water hoses, disconnect leads to lamps either side of grille panel and right-hand junction box. Remove fan blades, also bolts securing front apron and grille to cross-member and front wings. Take out radiator matrix and grille assembly. Disconnect and remove all pipes, wires and controls to engine unit.

Fit engine sling to cylinder head front and rear support brackets, and support gearbox with jack after removal of front floor and gearbox cover. Remove slave cylinder bracket from flywheel housing, take out bell housing nuts and washers, also mounting bolts. Manœuvre engine forward, up and clear of vehicle.

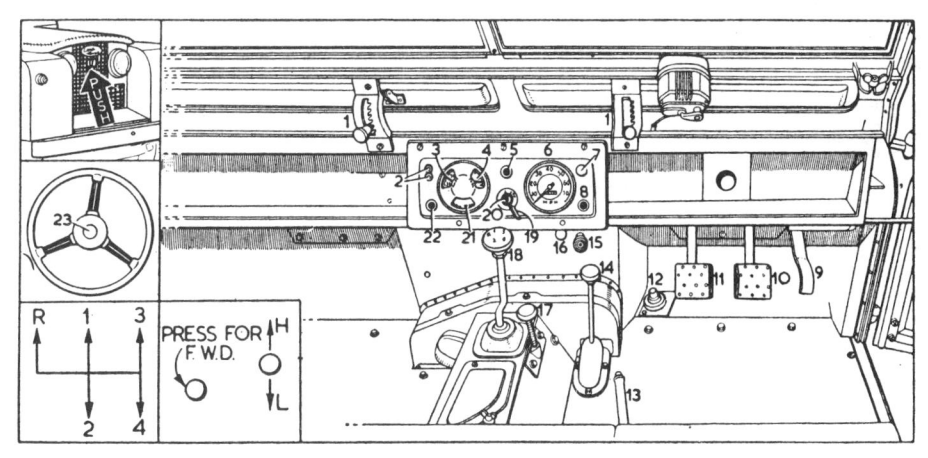

### INSTRUMENTS, CONTROLS, GEAR POSITIONS AND BONNET LOCK

1. Windscreen ventilators
2. Lead lamp sockets
3. Ammeter
4. Fuel gauge
5. Oil pressure warning light
6. Speedometer
7. Panel light switch
8. Cold start warning light
9. Accelerator
10. Brake pedal
11. Clutch pedal
12. Headlamp dipper switch
13. Handbrake
14. Transfer box lever
15. Starter switch
16. Cold start control
17. Front wheel drive control
18. Gear lever
19. Main lighting switch
20. Ignition switch
21. Main beam warning lamp
22. Ignition warning light
23. Horn push

Inserts show, top left: bonnet release details, below: siting of steering column mounted control, and bottom left, operative positions of gear lever and transfer box lever.

31

## ENGINE DATA

| General | |
|---|---|
| Type | o.h.v. |
| No. of cylinders | 4 |
| Bore x stroke: mm | 90.49 × 88.9 |
| in | 3.562 × 3.50 |
| Capacity: c.c. | 2286 |
| cu in | 139.5 |
| R.A.C. rated h.p. | 20.4 |
| Max. b.h.p. at r.p.m. | 77 @ 4250 |
| Max. torque at r.p.m. | 124 lb ft @ 2500 |
| Compression ratio | 7.0 : 1 |

## CRANKSHAFT AND CON. RODS

| | Main Bearings | Crankpins |
|---|---|---|
| Diameter | 2.50in | 2.126in |
| Length | 1.057-1.067in | — |

| | |
|---|---|
| Running clearance: | |
| main bearings | .001-.0025in |
| big ends | .0007-.0025in |
| End float: main bearings | .002-.006in |
| big ends | .007-.011in |
| Undersizes | .010, .020, .030, .040in |
| Con. rod centres | 6.904-6.908in |
| No. of teeth on starter ring gear pinion | 97/11 |

## PISTON AND RINGS

| | |
|---|---|
| Clearance (skirt) | .0023-.0028in |
| Oversizes | .010, .020, .030, .040in |
| Weight without rings or pin | 20¼ ozs |
| Gudgeon pin: | |
| diameter | .9998+.002in |
| fit in piston | Nil to .0002in interf. |
| fit in con. rod | .0003-.0005in |

| | Compression | Oil Control |
|---|---|---|
| No. of rings | 2 | 1 |
| Gap | .015-.020in | .015-.020in |
| Side clearance in grooves | .0005-.002in | .0005-.002in |
| Width of rings | .069-.070in | .185-.186in |

## CAMSHAFT

| | |
|---|---|
| Bearing journal: diameter | 1.842-1.843in |
| Bearing clearance | .001-.002in |
| End float | .0025-.0055in |
| Timing chain: pitch | ⅜in |
| No. of links | 78 |

## VALVES

| | Inlet | Exhaust |
|---|---|---|
| Head diameter | 1.755in | 1.380in |
| Stem diameter | .312in | .343in |
| Face-angle | 30° | 45° |

| | Inner | Outer | Inner | Outer |
|---|---|---|---|---|
| Spring length: | | | | |
| free | 1.61in | 1.76in | 1.61in | 1.76in |
| fitted | 1.38in | 1.50in | 1.37in | 1.49in |
| at load | 17.5lb | 46lb | 18.5lb | 48lb |

## NUT TIGHTENING TORQUE DATA

| | Bolt Size | lb/ft |
|---|---|---|
| **ENGINE** | | |
| Connecting rod bolts | — | 35 |
| Cylinder head nuts | ½in UNF | 75 |
| Main bearing bolts | 7/16 UNF | 85 |
| Rocker shaft support bracket bolts | 5/16 UNF | 12-13 |
| Flywheel securing bolts | — | 50 |
| **REAR AXLE** | | |
| Pinion flange nut | — | 85 |
| Crownwheel retaining bolts | | |
| (std. .375in) | — | 35 |
| (special .390in) | — | 45 |

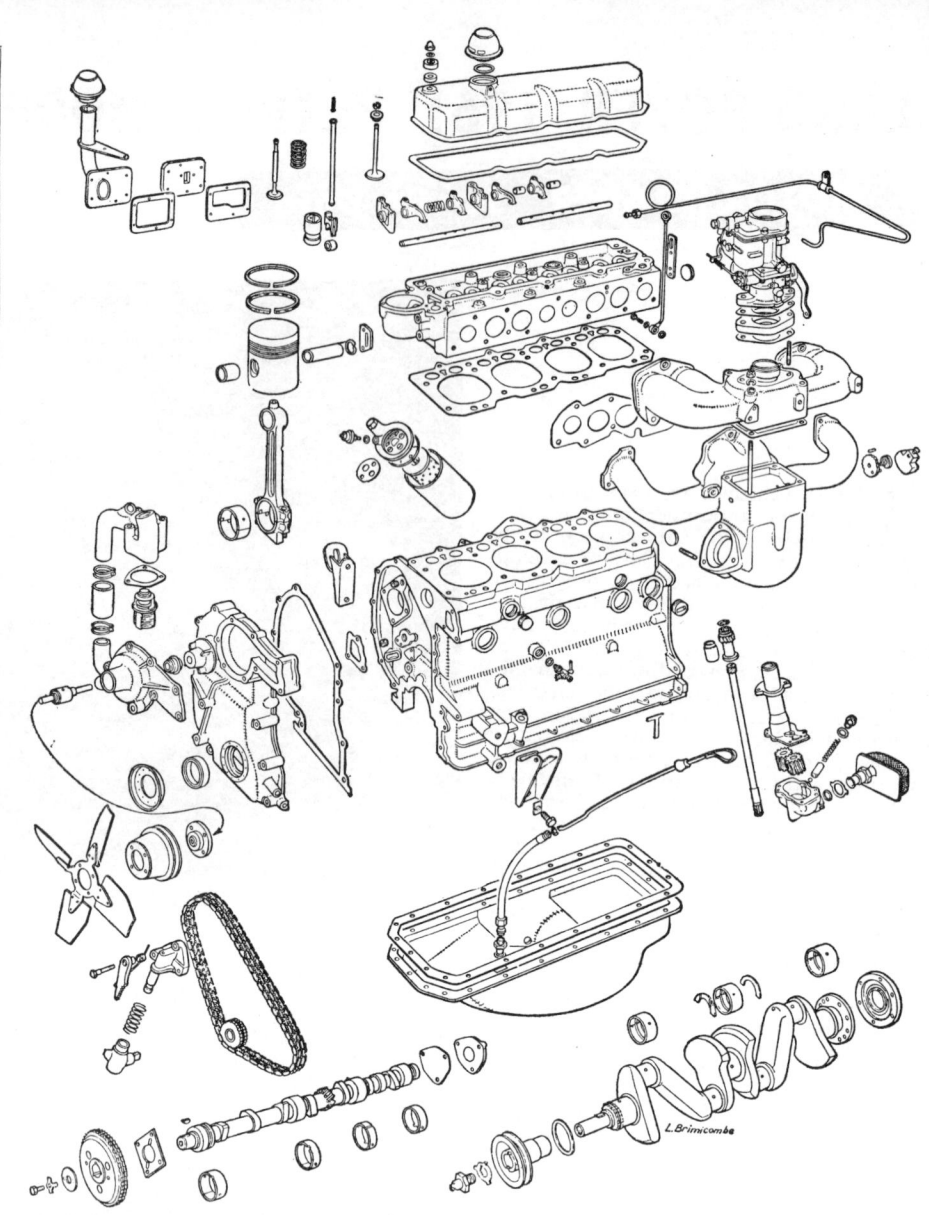

## SPECIAL TOOLS

| | Part No. |
|---|---|
| **ENGINE:** | |
| Chain wheel extractor | 507231 |
| Guide for rear main bearing cap seals | 270304 |
| Valve guide fitting tool | 274406 |
| Tappet guide extractor and fitting tool | 274397 |
| Camshaft bearing extractor | 274388 |
| Camshaft bearing fitting tool | 274381 |
| Starter dog nut spanner | 507234 |
| Camshaft bearing reamer | 274389 |
| **CLUTCH & GEARBOX** | |
| Gauge plate (release levers) | 277984 |
| Transfer box intermediate shaft extractor | 262772 |
| Gearbox mainshaft nut spanner | 263056 |
| **REAR AXLE** | |
| Diff. pinion rear bearing extractor | 262757 |
| Axle shaft retaining collar, removal and replacement tools | 275870 |
| Diff. pinion setting gauge | 262761 |
| Crown wheel locking nut "C" spanner | 262759 |
| **FRONT AXLE** | |
| Indicator bracket for hub adjustment | 272956 |
| Steel pin for steering relay assembly | 510309 |
| Drop arm extractor | 262776 |

Parts of the engine showing the fixed and moving components. Note assembly of the water pump parts, arrowed on the left of the illustration

## GENERAL DATA

| | |
|---|---|
| Basic vehicles: | |
| Wheelbase 88 | 7ft. 4in |
| 109 | 9ft 1in |
| Track: front | 4ft 3½in |
| rear | 4ft 3½in |
| Turning circle | 38ft |
| Turning circle 109 | 45ft |
| Ground clearance: | |
| 88 (6.00-16in tyres) | 8in |
| 88 (7.00-16in tyres) | 8½in |
| 109 (7.50-16in tyres) | 9¼in |
| Tyre size: front and rear 88 | 6.00-16 |
| 88 | 7.00-16 |
| 109 | 7.50-16 |
| Overall length 88 | 11ft 10¾in |
| Overall length 109 | 14ft 7½in |
| Overall width | 5ft 4in |
| Overall height 88 (hood up) | 6ft 5½in |
| 109 (top of cab) | 6ft 9in |
| Weight (dry) 88 | 2840lb |
| 109 | 3234lb |

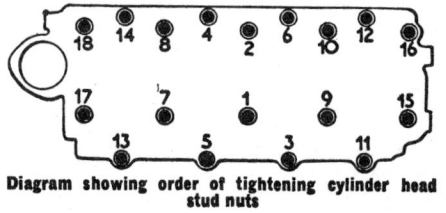

Diagram showing order of tightening cylinder head stud nuts

## Crankshaft

Three main bearings. Steel backed copper-lead-lined shell located in crankcase and caps. Endfloat controlled by spilt thrust washers recessed in block either side of centre main bearing. Fit with oil grooves towards housing. No hand fitting permissible, bearings may be changed without removal of shaft. Flywheel, with detachable starter ring gear spigoted on rear flange of crankshaft, dowel located and secured by setscrews. Self-lubricating spigot bush pressed into flywheel.

Timings sprocket keyed on front end of shaft, boss of sprocket outwards, with Woodruff key. Oil thrower disc trapped between sprocket and pulley hub. Assembly retained by started dog setscrew. Pulley hub passes through lipped oil seal (lip inwards) in timing cover.

Rear main bearing cap fits in square recess in crankcase with "T" shaped composition seals which fit between bearing cap and crankcase. Imperative to use oil seal guide tool No. 270304 secured to sump studs either side of rear drain bearing cap.

Split rubber moulding forming oil collector ring, dowel located on rear main bearing cap and block, and may be removed with the crankshaft *in situ*.

## Connecting Rods

"H"-section rod, big end split horizontally and small end bushed for fully floating gudgeon pin. Thin wall, steel-backed copper-lead-lined big end bearing shells located by tabs in rods and caps. No hand fitting permissible. Big end bolts located in con. rod shoulders and locked with self-locking nuts. Fit piston and rod assembly with bleed hole in rod away from camshaft.

## Pistons

Aluminium alloy. Fully floating gudgeon pins located by circlips. Fit of pin in piston is critical. Pin must not fall through by its own weight and must be fitted with hand pressure only.

Top compression ring chromium plated and of square section as is oil scraper ring, which latter is fitted above gudgeon pin lowest groove for use in service only. Second and third compression rings bevel edged and must be fitted with side marked "T" uppermost.

Big ends will pass through bores, remove and reassemble from top. Check ring gaps and side clearance together with piston fits in bores to dimensions in data tables.

## Camshaft

Duplex roller endless chain drive, with hydraulic tensioner.

Camshaft sprocket keyed on shaft with Woodruff key and retained by setscrew and lockwasher. Shaft runs in four split white metal-lined steel-backed bearings, notched for location in cylinder block. End float controlled by thrust plate trapped between sprocket and shoulder on shaft, bolted to crankcase.

Tensioner consists of an idler sprocket mounted on extension shaft of hydraulic cylinder; valve is contained within, and assembly is secured to crankcase casting by three setbolts and lockwashers. Locking pawl to secure lateral movement of tensioner is pivoted on piston securing bolt. Oil pressure from lubrication system augments spring, and oil is trapped by non-return valve in base of cylinder to give hydraulic lock.

To retime valves with timing chain and tensioner off, set exhaust tappets to running clearance, slacken inlet tappet screws right off and turn camshaft in running direction until No. 1 exhaust valve is fully open (use dial indicator if possible). Turn crankshaft in running direction until EP mark on flywheel is opposite pointer (visible under trap on off side of flywheel housing). Assemble timing chain so that there is no slack on driving side, and fit idler sprocket assembly. Check timing. Camshaft sprocket has three keyways for fine adjustment.

## Valves

Overhead, non-interchangeable, inlet larger than exhaust and thinner stem diameter. Split cone cotter fixing, double springs. Inner springs fit tightly in outer springs by selective assembly and should only be replaced as mated pairs. Sealing rings fitted on top of inlet valve guides.

Valve guides shouldered, not interchangeable, exhaust larger than inlet and of different bore diameter.

Press in guides from top of head, or use tool No. 274406 to pull guides into position with shoulders against machined bosses.

## Tappets and Rockers

Tappets and rockers are sliding fit in guides pressed into crankcase, retained and located by special hexagon-headed dowel bolts. Valves operated by pushrods, and rockers work on tubular steel two-stage shaft supported in five brackets mounted on cylinder head. Rockers are bushed and drilled for lubrication; oil is supplied to centre bracket from lead on gallery. Springs separate rockers for each cylinder and thrust washers are fitted between rockers and mounting brackets. When refitting ensure that $\frac{1}{2}$in UNF bolts which also secure cylinder head are tightened to 65lb/ft but $\frac{5}{16}$UNF bolts *only* to 12lb/ft.

## Lubrication

Gear pump in sump, spigoted in crankcase by integral drive housing and retained by two $\frac{5}{16}$in UNF setbolts and lockplates. Lower half of pump unit containing gears is bolted to upper casing by four $\frac{5}{16}$in UNF setbolts and dowelled for location. Gauze strainer screwed up to lower half of pump body, retained by large nut and lockwasher. Drive shaft, splined at lower end for engagement with pump driving gear and splines at top end engage internal splines of vertical drive shaft assembly which is bushed and unit is located by long grubscrew; circlip positioning drive shaft below lip of drive housing. Non-adjustable ball relief valve in lower half of pump housing, with spring and plunger inserted from outside and located by hexagon headed plug. Normal oil pressure is 50-60lb/sq in at 30 m.p.h. in top gear, engine hot.

## Cooling System

Pump, fan and non-adjustable thermostat in housing bolted to cylinder head above pump. Pump has spring-loaded carbon and rubber seal. Adjust fan belt by swinging dynamo until there is about $\frac{1}{16}$-$\frac{1}{4}$in movement either way on longest run of belt.

# TRANSMISSION

## Clutch

Borg & Beck single dry plate. Journal ball release bearing enclosed in separate housing bolted to gear box, from which it is lubricated, and operating clutch fingers through sliding sleeve.

Only adjustment is by rotation of slave cylinder rod to give $\frac{3}{4}$in free movement at pedal pad.

Gearbox can be removed for service to clutch without disturbing engine.

## Gearbox

Four-speed, synchromesh on top and 3rd gears, single helical constant mesh gears except for 1st and reverse. Two-speed transfer box bolted to rear.

**To remove gearbox and transfer box,** disconnect battery, remove hood (if fitted), detach centre panel from seat box, disconnect handbrake rod from bell-crank and remove lever assembly complete, drawing end of lever back through draft excluder. Unscrew knobs from gear lever, and transfer lever. Take up floor plates and remove gearbox cover. Remove seat box. Disconnect handbrake.

Remove rear propeller shaft complete, and disconnect front shaft and p.t.o. shaft (if fitted) at gearbox end. Remove clutch slave cylinder and lift off linkage, unhooking return spring; cross-shaft is jointed to extension shaft, supported in spherical bush on bracket. Extract either joint pin (split pinned) and remove bracket with extension shaft. Disconnect speedo drive and remove rear mounting bolts. Jack up rear of engine about $\frac{1}{2}$in, and take weight of gearbox on slings. Take off nuts and plain washers round bell-housing flange, draw gearbox back and lift out. Offside mounting bracket may have to be detached to clear.

**To remove transfer box** detach main gear lever assembly and reverse stop from gearbox and bell-housing. Take off transfer cover plate (with plug) on top of transfer box. Detach transfer gear lever with link from selector rod and bracket. Remove lever and detach

| BALL AND ROLLER BEARING DATA | | | |
|---|---|---|---|
| | Part No. | Int. dia., Ext. dia., Width (in or mm) | Type |
| **FRONT AXLE** | | | |
| Swivel pin bottom | 217268 | .750 × 2.125 × .875in | TR |
| Hub: inner | 217269 | 1.8125 × 3.3465 × .8125in | TR |
| outer | 217270 | 1.625 × 3.00 × .709in | TR |
| Differential as rear axle | | | |
| **GEARBOX** | | | |
| Primary shaft pinion | 55714 | 1.500 × 3.250 × .750in | B |
| Mainshaft: front | 06397 | 1.00 × 1.50 × 1.00in | B |
| rear | 1645 | .750 × 1.125 × 1.00in | B |
| Layshaft: front | 09962 | $\frac{3}{4}$ × 2 × $\frac{11}{16}$in | B |
| rear | 55715 | .874 × 2.00 × .565in | B |
| Clutch withdrawal | 214797 | 35 × 72 × 17 mm | B |
| **REAR AXLE** | | | |
| Crownwheel diff. bearing | 41045 | 1.500 × 3.00 × .9375in | TR |
| Bevel pinion (outer) | 219550 | 1.25 × 2.8593 × 1.1875in | TR |
| Bevel pinion (inner) | 219544 | 1.500 × 3.125 × 1.1563in | TR |
| Hub bearings (inner) | 217269 | 1.8125 × 3.3465 × .8125in | TR |
| Hub bearings (outer) | 217270 | 1.625 × 3.00 × .709in | TR |

bracket from bell-housing. Remove transmission brake drum and draw off rear driving flange. Detach brake back-plate assembly and bottom cover. Remove idler spindle locking plate and extract spindle out to rear with Special Tool No. 262772 catching idler gear. This gives access to three nuts inside, which, with six outside, hold transfer box to gearbox. Power take-off drive housing or blanking housing, with roller spigot bearing, should first be removed from back of transfer box.

**To dismantle gearbox** follow procedure in "Trader" Service Data sheet No. 150, covering Rover 60 and 75, which have similar gearbox.

**To reassemble gearbox** reverse order of dismantling, observing following:—

End float of 2nd and 3rd gears should be .004-.007in. Thrust washers .125 .128, .130 and .135in thick.

Conical distance-piece for front end of layshaft available in .312, .332 and .352in thicknesses to take up end float.

When inserting selector springs, note that 3rd/top (nearside) and 1st/2nd (centre) springs are same, but reverse (offside) is stronger.

Adjust 2nd speed stop screw so that with 2nd gear engaged there is .002in clearance between screw head and stop on selector rod.

## Transfer Box

Idler gear cluster runs on caged roller bearings on spindle, with thrust washers and shims at both ends, tabbed to locate in casing.

Output shaft carries constant mesh high gear (free) and sliding low gear (splined) between taper roller bearings, adjusted by shims (.005, .010, .015in thick) between casing and speedo drive housing.

Rear extension of output shaft carries speedo drive gear, nipped between inner race of rear bearing and driving flange, which also carries brake drum.

Forward extension of output shaft carries sliding dog for engaging four-wheel drive. When refitting output housing, cover end of selector rod with thimble to protect oil seal in housing.

**To dismantle transfer box** after removal from gearbox, with brake drum, driving flange, four-wheel drive gear off, remove speedo drive housing and shims, Pull off speedo drive gear. Tap output shaft back until outer race of rear bearing is free of casing. Extract spring ring retaining outer race of front bearing and tap output shaft forward as far as possible. Slide shaft back and insert aluminium packing pieces between rollers and outer race. Drive shaft forward again and repeat if necessary with thicker packings until outer race is free. Draw off inner race. Spring ring retaining high transfer gear on shaft with thrust washer can then be extracted, and shaft pushed out to rear through gears, which will drop out.

**To reassemble transfer box** reverse order of dismantling, observing following points:—

End float of high transfer gear on shaft should be .004-.008in after adjusting output shaft end float. Grind thrust washer if necessary.

## Propeller Shafts

Hardy Spicer needle roller bearing universal joints, series 1300 for both front and rear drive shafts and for p.t.o. shaft if fitted. Nipples for lubrication of all joints.

## Rear Axle

Fully floating spiral bevel drive. Rear cover welded to banjo casing.

To remove axle from car either drop rear ends of springs and roll axle back, or remove half shaft and brake assemblies and final drive and pass unit out sideways through road springs.

Differential unit details are similar to those of the current production cars. See Trader Service Data Nos. 227 and 297.

Bevel pinion shaft is carried in taper roller bearings, adjusted by shims.

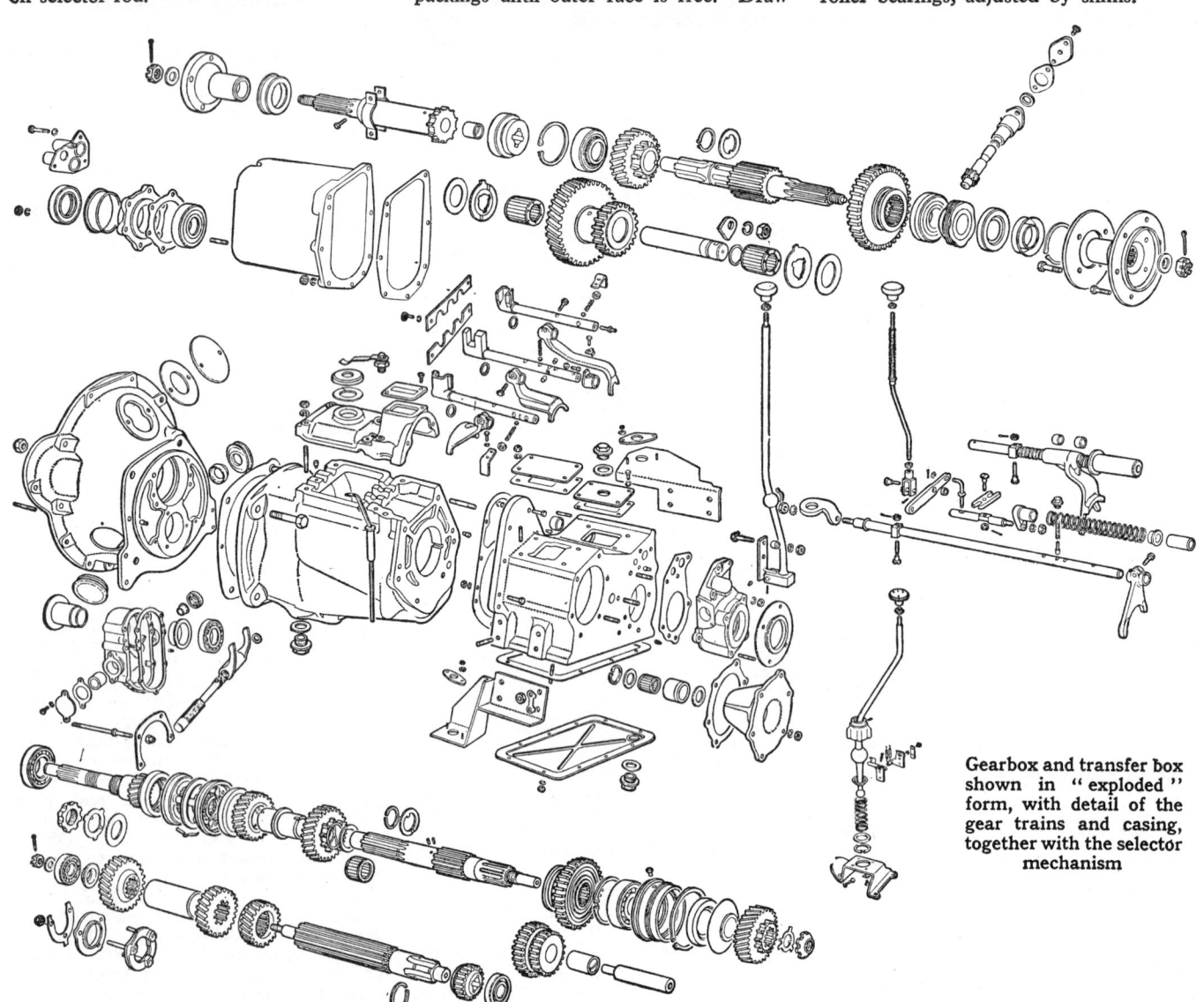

Gearbox and transfer box shown in "exploded" form, with detail of the gear trains and casing, together with the selector mechanism

# CHASSIS

## Brakes

Girling hydraulic 10in drums on 88in w.b. models with one snail cam adjuster on front and rear drums; 11in drums on 109in w.b. model; two snail cam adjusters on front drums and one square-ended adjuster on each rear drum. Handbrake operates Girling mechanical brake at rear of transfer box.

To adjust wheel brakes jack up wheel, turn snail cam adjuster until shoe binds, and back off until free.

To adjust handbrake and rear brakes of 109in model turn square-ended adjuster until shoes make contact with drum, and back off two clicks. Apply brake firmly to centralize shoes. To reset linkage after overhaul adjust hand lever end of pull rod *after adjusting shoes* so that lever pulls up two notches on ratchet before applying brake.

## Springs

Semi-elliptic front and rear. Bonded rubber shacklé and anchorage bushes. Plain bolts, tighten fully against inner member of rubber bush. Front springs are shackled at front, rear springs at rear. Front and rear bushes interchangeable, bushes in frame non-interchangeable.

## Front Axle

Final drive assembly interchangeable with rear axle. Inner swivel housings flange-bolted to ends of axle casing enclose Hardy Spicer universal joints. Driving members integral with half-shafts. Driven members integral with stub axle shafts, which are fully floating in hubs. Wheels run on taper roller bearings on stub axle tubes, which are flange-bolted to outer swivel housings.

Each inner swivel housing carries taper roller king pin bearings at lower end and spring-loaded cup and cone at upper end. Correct fit established when steering lever eye requires pull of 14-16lb to move it. Each half of two-piece king pin spigoted in outer swivel housing and registering in inner race of bearing. Shims (.003, .005, .010, .020in thick) under shoulder or each swivel pin for bearing adjustment to poundage figure given.

When assembling hub, adjust bearings to give .003-.004in end float.

Steering ball joints sealed side-plug type, pre-lubricated. Renew as assembly. Shanks threaded left- and right-hand, screwed into tubes and clamped. All six joints are identical except for thread of shank.

Adjustable lock stops have different settings for different tyres. Dimension from face of oil seal retainer to top of bolt head is $\frac{7}{16}$in for 6.00-16 tyres, $\frac{23}{32}$in for 7.00-16 tyres.

Steering relay lever and shaft assembly consists of tubular housing bolted vertically to cross-member, and carrying shaft in two split Tufnol conical bearings, which are heavily spring-loaded to damp steering.

To remove relay assembly detach grille, take out bolts holding grille panel to wings and chassis frame, and extract rubber washers under panel. Disconnect fore-and-aft drag link from upper lever, and draw off lower lever (cotter-clamped). Assembly can then be pushed out upwards after removal of two bolts.

To dismantle relay assembly detach upper lever, both end caps with oil seals, and brass thrust washers. Cover one end of shaft with heavy rag and tap shaft out, taking care as first split Tufnol bush is exposed, as spring is compressed to over 100lb. Release gently and tap shaft out with second bush. Keep pairs of bushes together. Spring data:—

No. of working coils 10
Free length ........ $7\frac{1}{4}$in
Fitted length ...... 3in
At load ............ $104\frac{1}{2}$lb

To reassemble relay fit top end plate and joint washer to housing. Fit one split bush to taper on bottom end of shaft, secure with 2in hose clip. Place steel washer on shaft, followed by spring. Insert prong of Tool No. 50323 through spring coils and through cross drilling in shaft and wind spring down tool until steel washer and Tufnol bush can be secured on taper at other end of shaft with hose clip. Place brass thrust washer on top end of shaft and insert into housing. Tap shaft home with plastic hammer until clips are freed, withdrawing tool as necessary. When assembly is complete and filled with oil, it should need at least 12lb on lever to turn shaft.

## Steering

Burman recirculating ball, cam supported at either end of box in cup and cone ball bearings. Shims for column end float adjustment provided under end plate; mesh adjustment by grubscrew and locknut in side cover. Adjust so that there is neither column end float nor rocker shaft end float.

## Shock Absorbers

Woodhead-Monroe telescopic. No attention needed.

## Power Take-off and Pulley

P.T.O. shaft at rear driven by propeller shaft from rear of gearbox through spur reduction gear in housing bolted to rear cross-member. Gears, 20 and 24 teeth, can be interchanged to give alternative ratios.

Pulley driven through spiral bevel gears in separate housing bolted to p.t.o. housing, and fitting over shaft splines.

To dismantle p.t.o. detach bearing caps and oil seal housings. Undo shaft nuts and tap shafts out of bearings and gears. Detach large bearing housings and remove gears. Both shafts run on taper roller bearings (all interchangeable). Outer races located in housing by spring rings. Inner races pulled up against gears by shaft nuts, with shims (.005, .010, .020in thick) for bearing adjustment so that they are free without play. Note that bolts on propeller shaft flange are retained by spring ring.

To dismantle pulley assembly draw off pulley with flange, and separate pulley shaft and bearing assembly from driving shaft housing (flange-bolted with shims for mesh). Tap shaft out of taper roller bearings.

Inner races of bearings separated by distance-piece with shims (.005, .010, .020in thick) for bearing adjustment.

Detach driving shaft end cap (flange-bolted with shims for bearing adjustment) with outer race of rear taper roller bearing. Inner race pressed on to hub of bevel pinion, which is retained on splined shaft by setscrew and large washer with cork washer behind. Shims between pinion and shoulder on shaft for mesh. Adjust all bearings to be free without play. Note that seal at outer end of pulley shaft is fitted with lip outwards to exclude dirt.

### CHASSIS DATA

**CLUTCH**

| | |
|---|---|
| Make ... ... ... ... | Borg & Beck |
| Type ... ... ... ... | s.d.p. |
| Springs: No. ... ... ... | 9 |
| colour ... ... ... | yellow/ lt. green |
| free length ... ... | 2.680in |
| Centre springs : no. ... ... | 3-buff/lt. green |
| colour ... ... | 3-white/ lt. green |
| Driven plate: | |
| thickness ... ... ... | .330in |
| dia. exit ... ... ... ... | 9in |
| max. permissible wear ... ... | .120in |

**GEARBOX**

| | | |
|---|---|---|
| No. of speeds ... ... | 4 | |
| Transfer box ... ... | 2 | |
| | High | Low |
| 1st ... ... ... ... | 16.171 | 40.688 |
| 2nd ... ... ... ... | 11.026 | 27.742 |
| 3rd ... ... ... ... | 7.435 | 18.707 |
| 4th ... ... ... ... | 5.396 | 13.578 |
| Rev. ... ... ... | 13.745 | 34.585 |

**PROPELLER SHAFT**

| | |
|---|---|
| Make ... ... ... | Hardy Spicer |
| Type ... ... ... | Needle roller bearing u.j. |

**FINAL DRIVE**

| | |
|---|---|
| Type ... ... ... | Fully floating s.b. |
| Crownwheel/bevel pinion teeth ... ... ... | 47/10 |

**BRAKES**

| | 88 | | 109 | |
|---|---|---|---|---|
| Type | Front | Rear | Front | Rear |
| Drum diameter | 10in | 10in | 11in | 11in |
| Lining: | | | | |
| length ... | $8\frac{1}{2}$in | $8\frac{1}{2}$in | 10.45in | 8.6in |
| width ... | $1\frac{1}{2}$in | $1\frac{1}{2}$in | $2\frac{1}{4}$in | $2\frac{1}{4}$in |
| thickness | $\frac{3}{16}$in | $\frac{3}{16}$in | $\frac{3}{16}$in | $\frac{3}{16}$in |

**SPRINGS**

| | Front | | Rear | |
|---|---|---|---|---|
| | N/S | O/S | N/S | O/S |
| Length (eye centres, flat) ... ... | $36\frac{1}{4}$in | $36\frac{1}{4}$in | 48in | 48in |
| Width ... ... | $2\frac{1}{2}$in | $2\frac{1}{2}$in | $2\frac{1}{2}$in | $2\frac{1}{2}$in |
| No. of leaves 88 ... | 9 | 9 | 11 | 11 |
| 109 ... | 11 | 11 | 10 | 10 |
| Free camber 88 ... | $5\frac{5}{16}$in | 6in | $6\frac{3}{4}$in | 7.42in |
| 109 ... | $5\frac{1}{8}$in | $5\frac{5}{8}$in | 8.2in | $9\frac{1}{4}$in |

**SHOCK ABSORBERS**

| | |
|---|---|
| Make ... ... ... ... | Woodhead-Monroe |
| Type ... ... ... ... | Telescopic double acting |
| Service ... ... ... ... | Replacement |

**STEERING BOX**

| | |
|---|---|
| Make ... ... ... ... | Burman |
| Type ... ... ... ... | Recirculating ball |
| Adjustments: | |
| column end float ... ... | shims |
| cross shaft end float / mesh ... ... | grubscrew & nut |

**FRONT-END SERVICE DATA**

| | |
|---|---|
| Castor ... ... ... ... | 3° |
| Camber ... ... ... ... | $1\frac{1}{2}$° |
| King pin inclination ... ... | 7° |
| Toe-in ... ... ... ... | $\frac{3}{64}$-$\frac{3}{32}$in |
| No. of turns lock to lock... ... | $3\frac{3}{4}$ |
| Adjustments: castor / camber ... | nil |
| toe-in ... ... | screwed track rod ends |

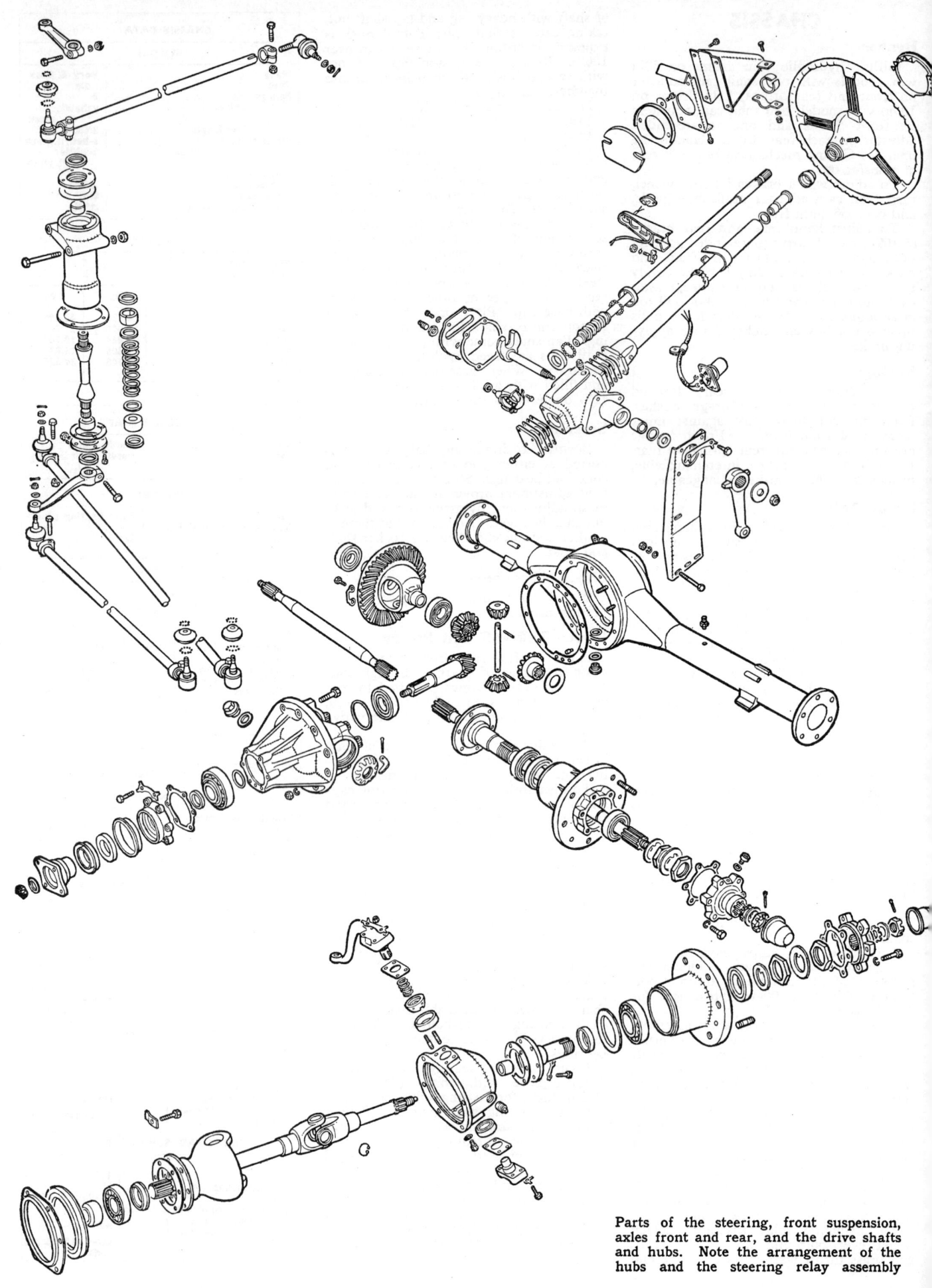

Parts of the steering, front suspension, axles front and rear, and the drive shafts and hubs. Note the arrangement of the hubs and the steering relay assembly

## TUNE-UP DATA

| | |
|---|---|
| Firing order | 1-3-4-2 |
| Tappet clearance (hot or cold): | |
| inlet | .010in |
| exhaust | .010in |
| Valve timing: | |
| inlet opens | 6° BTDC |
| inlet closes | 52° ABDC |
| exhaust opens | 34° BBDC |
| exhaust closes | 24° ATDC |
| Standard ignition timing | 6° BTDC (premium grade fuel) |
| Location of timing mark | Pointer on flywheel housing |
| Plugs: make* | Lodge |
| type | CLNH |
| size | 14mm |
| gap | .029–.032in |
| * also Champion N8 | |
| Carburettor: make | Solex d.d. |
| type | 40PA10-5A |
| Settings: Choke | 28mm |
| Main jet | 125 |
| Correction jet | 185 |
| Pilot jet | 50 |
| Pump jet | 65 |
| Economy jet | Blank |
| Air bleed jet | 1.5 |
| Starter petrol jet | 145 |
| Economy system petrol jet | 100 |
| Petrol level | ⅜ ± ⅛ in below float chamber joint face |
| Air cleaner: make | AC |
| type | Centrifugal, oil bath |
| Fuel pump: make | AC |
| type | Mech. |
| pressure | 1½–2½ lb/sq in |

## LUCAS EQUIPMENT

**BATTERY**
Model BT9A

**GENERATOR**
Model C39PV-2

**CONTROL BOX** — Part No. 22258E
Model RB106-2 — Part No. 37182H

**STARTING MOTOR**
Model M418G — Part No. 25533B
Drive: S-Type, Inboard (255194)

**DISTRIBUTOR**
Model DM2P4 — Part No. 40609
Max. centrifugal advance (crank degrees) 42° at 5,400 r.p.m.
No advance below 450 r.p.m.
Centrifugal advance springs — Part No. 55410187
Max. vacuum advance (crank degrees) 22°–26°
No advance below 2in Hg

**IGNITION COIL**
Model HA12 — Part No. 45054N
Primary resistance 3.0–3.5 ohms
Running current at 1,000 r.p.m. 1.0 amp

**WINDSCREEN WIPER**
Model FW2 — Part No. 75113

**HORN(S)**
Model WT618 — Part No. 69046F
(Low Note)
Type: Windtone
Current consumption 7.5–8.5 amp

**FLASHER UNIT**
Model FL5 — Part No. 35010A

**FUSE UNIT**
Model SF5-2
Fuse rating 35 amp (188218)

| Lamps | Model | Part No. | Bulb No. | Wattage | Cap |
|---|---|---|---|---|---|
| Head: R.H.D. | F700 | 51780 | 414 | 50/40 | B.P.F. |
| L.H.D. | F700 | 51533 | 355 | 42/36 | B.P.F. |
| N.A.D.A. | F700 | 51467 | — | — | — |
| (later) | — | 58563 | — | — | — |
| EUROPE (except France and Sweden) | F700 | 58286 | 410 | 45/40 | "Unified" |
| FRANCE | F700EF | 58287 | 411 | 45/40 | "Unified" |
| SWEDEN | F700 | 58459 | 410 | 45/40 | "Unified" |

| Lamps | Model | Part No. |
|---|---|---|
| Side | 638 | 52437 |
| Front flasher | 639 | 52438 |
| Stop tail | 581 | 53783 |
| Rear flasher | 637 | 53806 |
| Number plate | 467-2 | 53876 |
| Ignition warning | WL3-1 | 38046 |
| Flasher warning | WL13 | 38084 |
| Choke warning | WL3-1 | 38043 |
| Oil warning | WL3-1 | 38018 |

## DRAINING POINTS

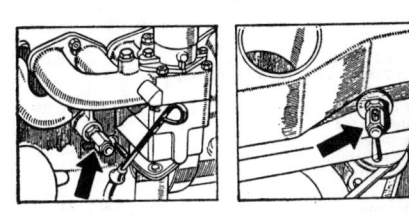

Left: shows the cylinder block draining point adjacent to the dipstick, beneath the manifolds. Right: the radiator matrix drain point, access from beneath

## LUCAS SWITCHES

| | Model | Part No. |
|---|---|---|
| Starter | ST18 | 076033 |
| Lighting and ignition | PR83 | 31270 or 031421 (N.A.D.A.) |
| Direction indicator | PR87 | 31509 (knob 312619) |
| Dip | FS22 or 218A (L.H.D.) | 31372 31800A |
| Stop light | HL2 | 31082 |
| Horn push | HP19 | 76205 |
| Heater | 3R | 78356 (knob 54332086) |
| Choke | 54G | 31540A |

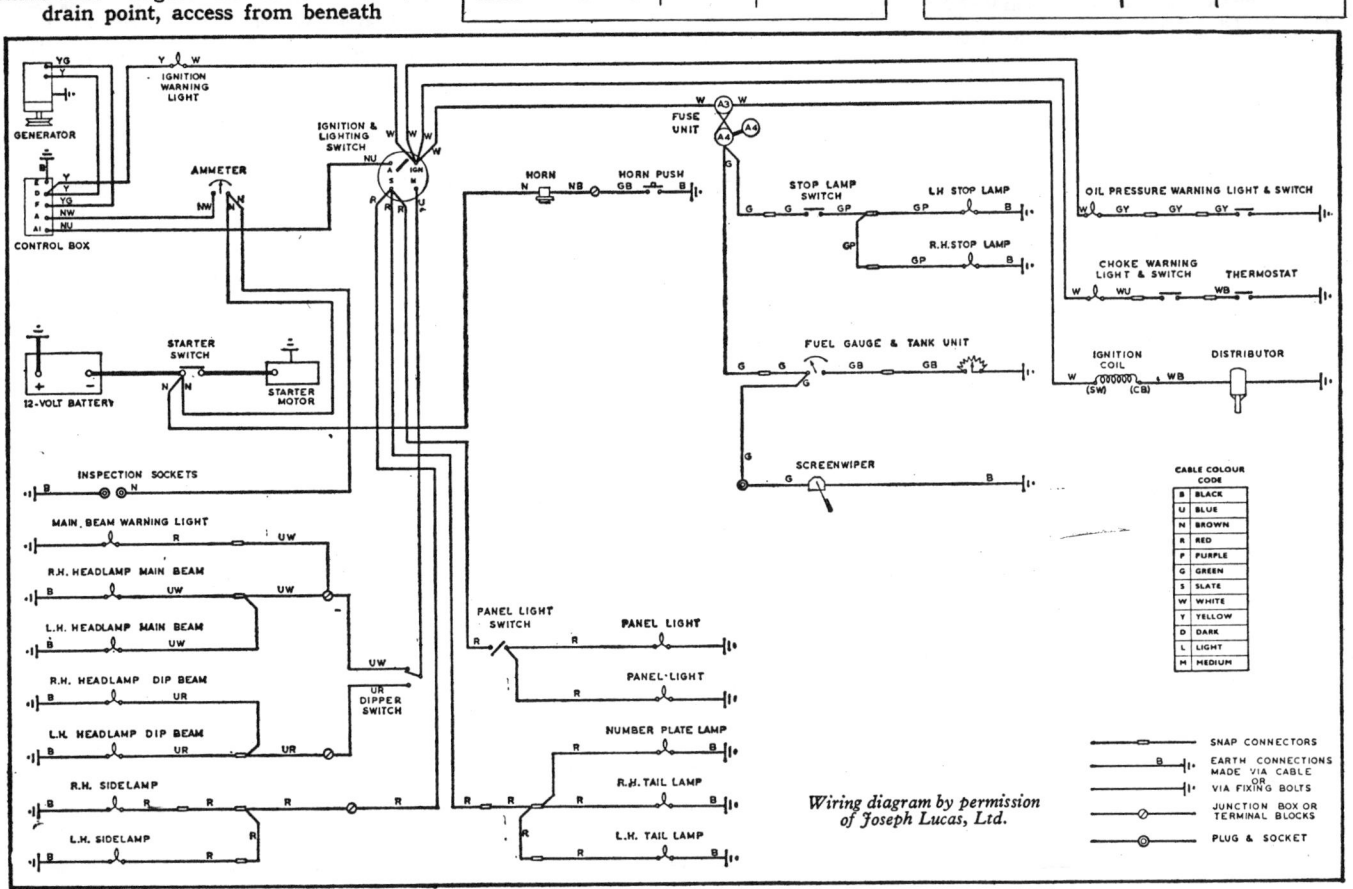

## SUNDRY EQUIPMENT

| | Model | Part No. |
|---|---|---|
| Mirror | 434/108 | 62054 |
| Ammeter | CZU30 | 36159 |
| Inspection sockets | — | 39517 |
| Junction box | 6J | 78266 |

*Wiring diagram by permission of Joseph Lucas, Ltd.*

**CABLE COLOUR CODE**

| | |
|---|---|
| B | BLACK |
| U | BLUE |
| N | BROWN |
| R | RED |
| P | PURPLE |
| G | GREEN |
| S | SLATE |
| W | WHITE |
| Y | YELLOW |
| D | DARK |
| L | LIGHT |
| M | MEDIUM |

— SNAP CONNECTORS
— EARTH CONNECTIONS MADE VIA CABLE OR VIA FIXING BOLTS
— JUNCTION BOX OR TERMINAL BLOCKS
— PLUG & SOCKET

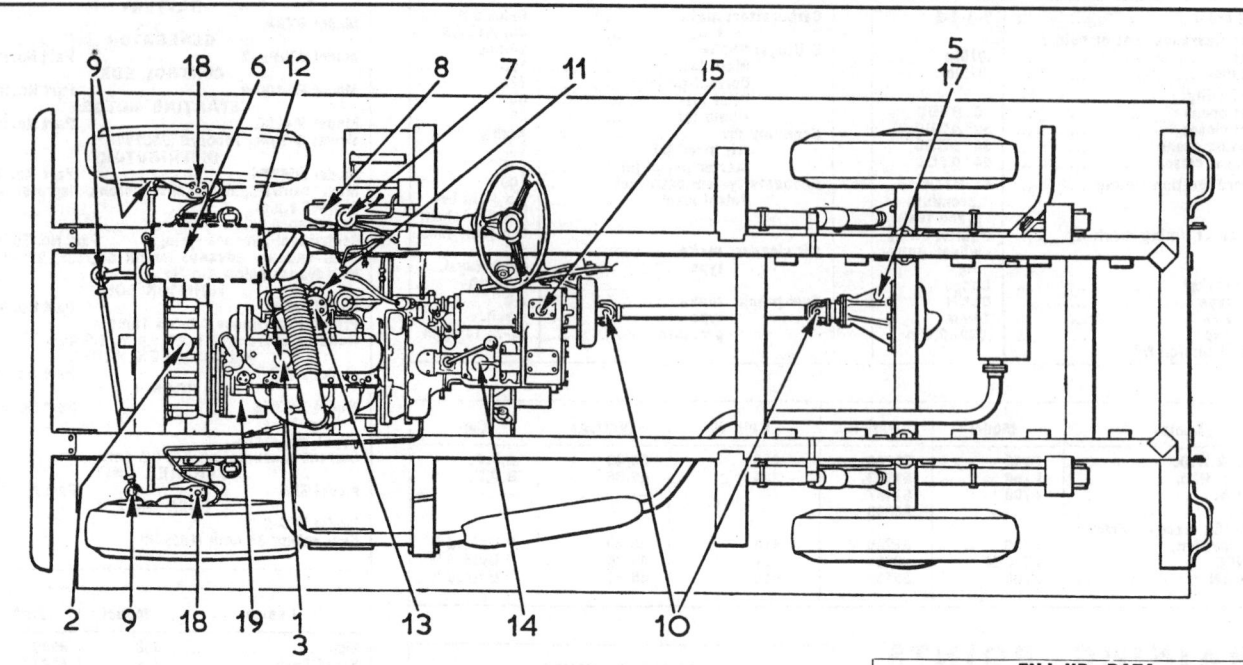

## KEY TO MAINTENANCE DIAGRAM

1. Engine sump } check and top up
2. Radiator

**EVERY 3,000 MILES (or 120 hrs.)**
3. Engine sump—drain and refill
4. Front axle
5. Rear axle } check and top up
6. Battery
7. Clutch and brake fluid reservoir
8. Steering box
9. Steering joints—check rubber boots
10. Propeller shafts—grease

**EVERY 6,000 MILES (or 240 hrs.) as for 3,000 miles plus following:**
11. Engine oil external filter element—renew
12. Breather filters—clean
13. Distributor—oil shaft bearing, contact breaker pivot and auto advance mechanism, smear cam with grease

**EVERY 9,000 MILES (or 360 hrs.) as for 6,000 miles plus following:**
14. Gearbox
15. Transfer box
16. Front axle } drain and refill
17. Rear axle
18. Front swivel pin housings

**EVERY 12,000 MILES (or 480 hrs.) as for 9,000 miles plus following:**
19. Dynamo—lubricate

### FILL-UP DATA

|  | Pints | Litres |
|---|---|---|
| Engine sump | 11 | 6.0 |
| Extra when refilling after fitting new filter | 3 | 1.75 |
| Air cleaner | 1½ | 0.85 |
| Main gearbox | 2½ | 1.5 |
| Transfer box | 4½ | 2.5 |
| Rear differential | 3 | 1.75 |
| Front differential | 3 | 1.75 |
| Swivel pin housing (each) | 1 | 0.5 |
| Fuel tank | 10 galls | 45 |
| Hydraulic front winch, supply tank | 4½ galls | 20.0 |
| Hydraulic front winch, gearbox | 2 | 1.0 |
| †Tyre pressures, front and rear (normal 88 model) | 25lb/sq in | 1.75 Kg/cm² |
| front and rear (fully laden) | 30lb/sq in | 2.1Kg/cm² |
| †109 model (normal) | 22lb/sq in | 1.55Kg/cm² |
| †109 model (full load) rear | 36lb/sq in | 2.53Kg/cm² |

## RECOMMENDED LUBRICANTS

| COMPONENTS | S.A.E. | B.P. Energol | CASTROL | DUCKHAM'S | ESSO | MOBIL | REGENT | SHELL |
|---|---|---|---|---|---|---|---|---|
| Engine, Air Cleaner and Governor | 20W | SAE 20W | Castrolite | NOL Twenty | Extra Motor Oil 20W/30 | Mobiloil Arctic | Advanced Havoline 20/20W | Shell X100 SAE 20/20W |
| Gearbox and Transfer Box | 90EP | EP SAE 90 | Hypoy | Hypoid 90 | Gear Oil GP 90 | Mobilube GX 90 | Universal Thuban 90 | Spirax 90 EP |
| Differentials and Swivel Pin Housings | 90EP | EP SAE 90 | Hypoy | Hypoid 90 | Gear Oil GP 90 | Mobilube GX 90 | Universal Thuban 90 | Spirax 90 |
| Steering Box | 90EP | EP SAE 90 | Hypoy | Hypoid 90 | Gear Oil GP 90 | Mobilube GX 90 | Universal Thuban 90 | Spirax 90 EP |
| Steering Relay Unit (sealed) | 90EP | EP SAE 90 | Hypoy | Hypoid 90 | Gear Oil GP 90 | Mobilube GX 90 | Universal Thuban 90 | Spirax 90 EP |
| Rear Power Take-off, Pulley Unit and Capstan Winch | 90EP | EP SAE 90 | Hypoy | Hypoid 90 | Gear Oil GP 90 | Mobilube GX 90 | Universal Thuban 90 | Spirax 90 EP |
| Hydraulic Winch Gearbox | 90EP | EP SAE 90 | Hypoy | Hypoid 90 | Gear Oil GP 90 | Mobilube GX 90 | Universal Thuban 90 | Spirax 90 EP |
| Hydraulic Winch Supply Tank | — | — | Hyspin 70 or Castrolite | — | Teresso 43 or Essolube HD10W | DTE Light | — | Tellus 27 |
| Lubrication Nipples | — | Energrease L2 | Castrolease LM | LB10 Grease | Multi-purpose Grease H | Mobilgrease MP | Marfak Multi-purpose 2 | Retinax A |

Note: These recommendations apply to temperate climates where operational temperatures may vary between approx. 10°F. (12°C.) and 90°F. (32°C.). Information on oil recommendations for use under extreme winter or tropical conditions may be obtained through Rover dealers or distributors or from the Technical Service Dept., Rover Co., Ltd. Multigrade oils of a 10W/30 rating produced by the companies listed above are also recommended for the engine, subject to its being in good mechanical condition.

# LAND-ROVER SERIES II (DIESEL)

| ENGINE DATA | |
|---|---|
| Type ... ... ... ... | Diesel |
| No. of cylinders ... ... ... | 4 |
| Bore × stroke: mm ... ... | 85.725 × 88.9 |
| in ... ... | 3.375 × 3.50 |
| Capacity: cc.. ... ... ... | 2052 |
| cu. in ... ... ... | 125.67 |
| R.A.C. rated h.p. ... ... ... | 18.3 |
| Max. b.h.p. at r.p.m. ... ... | 52 @ 3500 |
| Max. torque at r.p.m. ... ... | 87 lb/ft @ 2000 |
| Compression ratio ... ... | 22.5 : 1 |

| CRANKSHAFT AND CON. RODS | | | | |
|---|---|---|---|---|
| | Main Bearings | | | Crankpins |
| Diameter ... | 2.500 + .000in − .001in | | | 2.126 + .000in − .001in |
| | Front | Centre | Rear | |
| Length | 1.239 + .002in | 1.310 + .002in | 1.955 − .008in | 1.310 + .002in |

| | |
|---|---|
| Running clearance: | |
| main bearings ... ... ... | .001-.0025in |
| big ends ... ... ... | .001-.0025in. |
| End float: | |
| main bearings ... ... ... | .002-.006in |
| big ends ... ... ... | .007-.011in. |
| Undersizes ... ... ... | .010, .020, .030, .040in |
| Con. rod centres ... ... ... | 6.906 ± .002in |
| No. of teeth on starter ring gear/pinion ... ... ... ... | 100/10 |

| PISTONS AND RINGS | |
|---|---|
| Clearance (skirt) ... ... | .004–.005in |
| Oversizes ... ... ... | nil |
| Gudgeon pin: diameter ... | 1.187 + .002in |
| fit in piston ... | zero–.0002 interf. |
| fit in con. rod ... | .0003–.0005in clear |

| | Compression | Oil Control |
|---|---|---|
| No. of rings | 3 | 1 |
| Gap ... | .010–.015in | .010–.015in |
| Side clearance in grooves | .0025–.0035in | .0025–.0035in |
| Width of rings | .092–.093in | .186–.187in |

| CAMSHAFT | |
|---|---|
| Bearing journal: diameter ... | 1.841–1.842in |
| Bearing clearance ... ... | .001–.002in |
| End float ... ... ... | .0025–.0055in |
| Timing chain: pitch ... ... | $\frac{3}{8}$ |
| no. of links ... | 78 |

| VALVES | Inlet | | Exhaust | |
|---|---|---|---|---|
| Head diameter | 1.1546–1.1542in | | 1.316–1.312in | |
| Stem diameter | .310–.311in | | .342–.343in | |
| Face-angle ... | 45° + $\frac{1}{4}$ | | 45° + $\frac{1}{4}$ | |
| | Inner | Outer | Inner | Outer |
| Spring length: | | | | |
| free in ... | 1.61 | 1.76 | 1.61 | 1.76 |
| fitted in ... | 1.383 | 1.50 | 1.37 | 1.497 |
| at load ... | 17.5lb | 46lb | 18.5lb | 48lb |

| NUT TIGHTENING TORQUE DATA | Bolt Size | lb/ft |
|---|---|---|
| ENGINE | | |
| Connecting rod bolts ... ... | — | 35 |
| Cylinder head nuts ... ... | $\frac{1}{2}$in UNF | 75 |
| Main bearing bolts ... ... | $\frac{9}{16}$in UNF | 85 |
| Rocker shaft support bracket bolts ... ... ... | $\frac{5}{16}$in UNF | 12-13 |
| Flywheel securing bolts ... | — | 50 |
| REAR AXLE | | |
| Pinion flange nut ... ... | — | 85 |
| Crownwheel retaining bolts | | |
| (std. .375in) ... ... | — | 35 |
| (special .390in) ... ... | — | 45 |

## ENGINE

### Removal

Remove engine without gearbox. Disconnect and take off bonnet, drain coolant, remove water hoses, disconnect leads to lamps either side of grille panel and right-hand junction box. Remove four blades, also bolts securing front apron and grille to cross-member and front wings. Take out radiator matrix and grille assembly. Disconnect and remove all pipes, wires and controls to engine unit.

Fit engine sling to cylinder head front and rear support brackets; disconnect clutch slave cylinder hose at bracket on dash and support gearbox with jack after removal of front floor and gearbox cover. Remove slave cylinder bracket from flywheel housing, take out bell housing nuts and washers also mounting bolts. Manœuvre engine forward to clear gearbox primary shaft and up and outwards clear of vehicle.

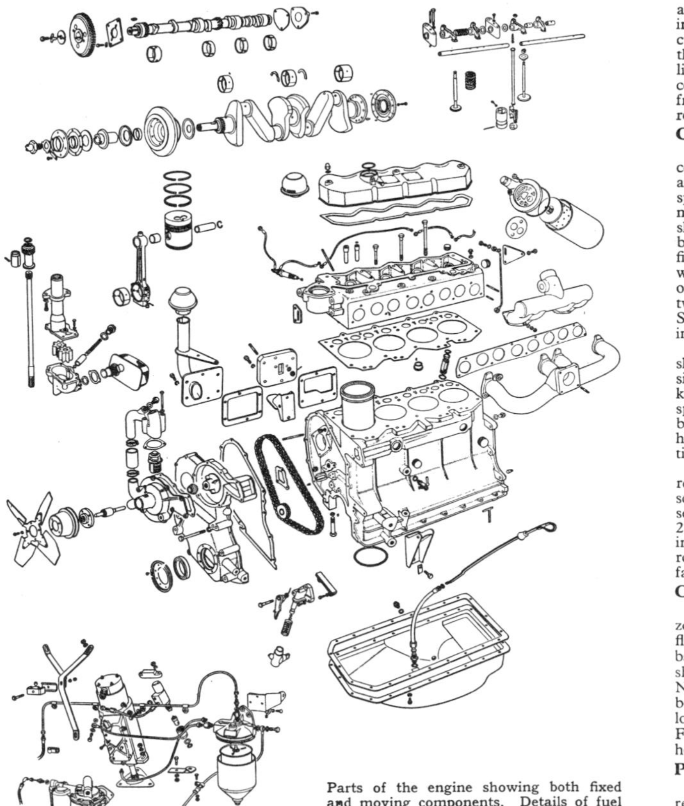

Parts of the engine showing both fixed and moving components. Details of fuel injection system and fuel filter are shown bottom left

| SPECIAL TOOLS | Part No. |
|---|---|
| DIESEL ENGINE | |
| Injector nozzle testing and setting outfit | 271483 |
| Safety cam ... ... ... ... | 276278 |
| Flushing tool—injector nozzle... ... | 278181 |
| Adaptor for Pintaux injector nozzle ... | 278182 |
| Injector nozzle cleaning outfit ... | 271484 |
| Valve guide fitting tool ... ... ... | 274406 |
| Push rod tube fitting tool ... ... | 274402 |
| Camshaft bearing extractor ... ... | 274388 |

N.B. For special tools for axle dismantling and assembly, see Trader Service Data No. 227, p. iii

| GENERAL DATA | |
|---|---|
| Wheelbase: 88in ... ... ... | 7ft 4in |
| 109in ... ... ... | 9ft 1in |
| Track: front ... ... ... ... | 4ft $3\frac{1}{2}$in |
| rear ... ... ... ... | 4ft $3\frac{1}{2}$in |
| Turning circle: 88 ... ... ... | 38ft |
| 109 ... ... ... | 45ft |
| Ground clearance: | |
| 88 (6.00—16in tyres) ... ... | 8in |
| 88 (7.00—16in tyres) ... ... | $8\frac{3}{4}$in |
| 109 (7.50—16in tyres) ... ... | $9\frac{3}{4}$in |
| Tyre size—front and rear: 88 ... | 6.00—16in |
| 88 ... | 7.00—16in |
| 109 ... | 7.50—16in |
| Overall length: 88 ... ... ... | 11ft $10\frac{3}{8}$in |
| 109 ... ... ... | 14ft 7in |
| Overall width ... ... ... ... | 5ft 4in |
| Overall height: 88 (hood up) ... | 6ft $5\frac{1}{2}$in |
| 109 (top of cab) ... | 6ft 9in |
| Weight (dry): 88 ... ... ... | 3,037 lb |
| 109 ... ... ... | 3,391 lb |

### Cylinders and Head

Cylinder block and crankcase is a one-piece casting, fitted with wet detachable type cast iron liners located in block by flange at top of liner. Water seal is effected by a flange at the top of the liner, and at the bottom by rubber rings located in grooves either side of water hole in cylinder block, and compressed against the side of the liner. To remove the liners, remove cylinder head, pistons and connecting rods and withdraw liners from block. Renew sealing rings on re-assembly.

### Crankshaft

Three main bearings. Steel-backed, copper-lead-lined shells dowelled in caps and crankcase. End float controlled by split thrust washers either side of centre main bearing; fit with grooves to crankshaft. No hand fitting permissible on bearings. Tighten cap nuts to torque figure specified in data tables. Flywheel with integral starter ring gear, spigoted on rear flange of crankshaft, located by two dowels and retained by eight setbolts. Spigot bush for primary shaft pressed into flywheel.

Timing sprocket keyed to front end of shaft with combined fan pulley and torsional vibration damper by single feather key. Oil thrower trapped between sprocket and pulley. Assembly retained by hand starter dog setscrew. Pulley hub passes through lipped oil seal in timing cover.

Rear main bearing cap fits in square recess in crankcase with "T" shaped square section cork seals in groove. Fit seals in cap using Special Tool No. 270304 to facilitate seal location with cap insertion. Split oil collector housing fits round rear of shaft and is bolted to rear face of crankcase and cap.

### Connecting Rods

"H"-section rod, big end split horizontally and small end bushed for fully floating gudgeon pin. Thin wall, steel-backed copper-lead-lined big end bearing shells located by tabs in rods and caps. No hand fitting permissible. Big end bolts located in con. rod shoulders and locked with split pins and slotted nuts. Fit piston and rod assembly with bleed hole in rod away from camshaft.

### Pistons

Aluminium alloy, combustion and swirl recess in crown, solid skirt. Fully floating gudgeon pins located by circlips. Fit of pin in piston is critical. Pin must not fall through by its own weight and must be fitted with hand pressure only.

Top compression ring chromium plated and of square section as is oil scraper ring, which latter is fitted above gudgeon pin lowest groove for use in service only. Second and third compression rings bevel edged and must be fitted with side marked "T" uppermost.

| BALL AND ROLLER BEARING DATA | Part No. | Int. dia., Ext. dia., Width (in or mm) | Type |
|---|---|---|---|
| FRONT AXLE | | | |
| Swivel pin bottom ... ... | 217268 | .750 × 2.125 × .875in | TR |
| Hub: inner ... ... ... | 217269 | 1.8125 × 3.3465 × .8125in | TR |
| outer ... ... ... | 217270 | 1.625 × 3.00 × .709in | TR |
| Differential as rear axle | | | |
| GEARBOX | | | |
| Primary shaft pinion ... ... | 55714 | 1.500 × 3.250 × .750in. | B |
| Mainshaft: front ... ... | 06397 | 1.00 × 1.50 × 1.00in | B |
| rear ... ... .. | 217478 | .750 × 1.125 × 1.00in | B |
| Layshaft: front ... ... ... | 09962 | $\frac{3}{4}$ × 2 × $\frac{11}{16}$in | B |
| rear ... ... ... | 55715 | .874 × 2.00 × .565in | B |
| Clutch withdrawal ... ... | 214797 | 35 × 72 × 17 mm | B |
| REAR AXLE | | | |
| Crownwheel diff. bearing ... | 41045 | 1.500 × 3.00 × .9375in | TR |
| Bevel pinion (outer) ... ... | 219550 | 1.25 × 2.8593 × 1.1875in | TR |
| Bevel pinion (inner) ... ... | 219544 | 1.500 × 3.125 × 1.1563in | TR |
| Hub bearings (inner) ... ... | 217269 | 1.8125 × 3.3465 × .8125in | TR |
| Hub bearings (outer) ... ... | 21720 | 1.625 × 3.00 × .709in | TR |

Big ends will pass through bores, remove and reassemble from top. Check ring gaps and side clearance together with piston fits in bores to dimensions in data tables.

### Camshaft

Duplex roller endless chain drive, with hydraulic tensioner.

Camshaft sprocket keyed on shaft with Woodruff key and retained by setscrew and lockwasher. Shaft runs in four split white metal-lined steel-backed bearings, notched for location in cylinder block. End float controlled by thrust plate trapped between sprocket and shoulder on shaft, bolted to crankcase.

Tensioner consists of an idler sprocket mounted on extension shaft of hydraulic cylinder; valve is contained within, and assembly is secured to crankcase casting by three setbolts and lockwashers. Locking pawl to secure lateral movement of tensioner is pivoted on piston securing bolt. Oil pressure from lubrication system augments spring, and oil is trapped by non-return valve in base of cylinder to give hydraulic lock. Anti-rattle pad is fitted on short side of timing chain and consists of composition type material bonded to steel plate and flange bolted to crankcase.

To retime valves with timing chain and tensioner off, set exhaust tappets to running clearance, slacken inlet tappet screws right off and turn camshaft in running direction until No. 1 exhaust valve is fully open (use dial indicator if possible). Turn crankshaft in running direction until EP mark on flywheel is opposite pointer (visible under trap on off side of flywheel housing). Assemble timing chain so that there is no slack on driving side, and fit idler sprocket assembly. Check timing. Camshaft sprocket has three keyways for fine adjustment.

### Valves

Overhead, non-interchangeable, inlet larger than exhaust and thinner stem diameter. Split cone cotter fixing, double springs. Inner springs fit tightly in outer springs by selective assembly and should only be replaced as mated pairs. Sealing rings fitted on top of inlet valve guides.

Valve guides shouldered, not interchangeable, exhaust larger than inlet and of different bore diameter.

Press in guides from top of head, or use tool No. 274406 to pull guides into position with shoulders against machined bosses.

### Tappets and Rockers

Tappets and rockers are sliding fit in guides pressed into crankcase, retained and located by special hexagon headed dowel bolts. Valves operated by pushrods, and rockers work on tubular steel two-stage shaft supported in five brackets mounted on cylinder head. Rockers are bushed and drilled for lubrication; oil is supplied to centre bracket from lead on gallery. Springs separate rockers for each cylinder and thrust washers are fitted between rockers and mounting brackets. When refitting ensure that $\frac{1}{2}$in UNF bolts which also secure cylinder head are tightened to 75lb/ft but $\frac{5}{16}$in UNF bolts securing rocker brackets *only* be tightened to 12lb/ft.

### Lubrication

Gear pump in sump, spigoted in crankcase by integral drive housing and retained by two $\frac{5}{16}$in UNF setbolts and lockplates. Lower half of pump unit containing gears is bolted to upper casing by four $\frac{5}{16}$in UNF setbolts and dowelled for location. Gauze strainer screwed up to lower half of pump body, retained by large nut and lockwasher. Drive shaft, splined at lower end for engagement with pump driving gear and splines at top end engage internal splines of vertical drive shaft assembly which is bushed and unit is located by long grubscrew; circlip positioning drive shaft below lip of drive housing. Non-adjustable ball relief valve in lower half of pump housing, with spring and plunger inserted from outside and located by hexagon headed plug. Normal oil pressure is 50-60lb/sq in at 30 m.p.h. in top gear.

### Fuel System

A distributor self-governing unit is fitted, together with Pintaux nozzles of C.A.V. manufacture. Setting figures are given in the data tables on p. vii.

### Cooling System

Pump, fan and non-adjustable thermostat in housing bolted to cylinder head above pump. Pump has spring-loaded carbon and rubber seal. Adjust fan belt by swinging dynamo until there is about $\frac{3}{16}$–$\frac{1}{4}$in movement either way on longest run of belt.

| Sundry Equipment | Model | Part No. |
|---|---|---|
| Mirror ... ... ... | 434/108 | 62054 |
| Ammeter ... ... ... | CZU34 | 36159 |
| Inspection sockets... ... | — | 39517 |
| Junction box ... ... | 6J | 78266 |
| Trafficator ... ... ... | SE100 | 54050 |
| Trailer socket ... ... | TLC4 | 82046 |
| Trailer plug and lead ... | — | 82047 |
| Flasher plug and lead ... | TLC4 | 82045 |
| Dummy socket ... ... | — | 809121 |
| **C.A.V. FUEL INJECTION EQUIPMENT** | | |
| Distributor fuel pump ... | DPA | 3240091 |
| Nozzle holder ... ... | — | BKB40 5008D |
| Nozzle ... ... ... | — | BDNOSP 6209 |
| Filter ... ... ... | F | F4-2 |

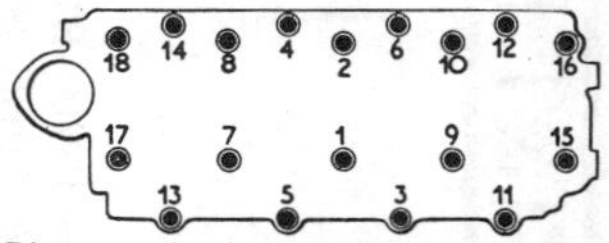


Diagram showing order of tightening cylinder head stud nuts. See also table of "Nut Tightening Torque Data."

## CT&T ROAD TEST

# Land Rover

IN TWO RECENT issues (March and April, 1961) Canada Track and Traffic printed the story, under the title "Drive into Danger", of the first vehicle to be driven overland from North to South America.

No road test ever called for a more punishing trial than the dense, trackless jungles of the Darien, the giant gullies, precipitous embankments, swamps, and, where there were roads, washouts and huge potholes. The vehicle chosen for the expedition was a Land-Rover, a squared-off rugged machine which has been referred to more than once as the "Jeep with a British accent."

The description is more pat than apt, for the Land-Rover, built by the Rover Company Ltd., of Warwickshire, England, is designed to do more, carry more and endure more than its U.S. counterpart.

"Journey into Danger" naturally dealt with all aspects of the expedition but the exploits of the Land-Rover as it crunched its way south attracted sufficient interest from readers to warrant a closer look at this remarkable vehicle.

The Land-Rover is sold in Canada in eight versions, canvas top, truck cab, hard top and station wagon, each model being available on 88 and 109-inch wheelbases. Canada Track and Traffic through the co-operation of the Rover Motor Company of North America, covered the 1961 Canadian Winter Rally from the comfort of a hardtop Land-Rover which proved fully equal to all our needs.

For the purposes of a report on the Land-Rover CT & T borrowed a short wheelbase station wagon and drove it along 401 Highway to Mosport Park where it was thrashed up hill and down dale through mud and mire.

All Land-Rovers consist primarily of a front compartment which will seat the driver and two passengers on a transverse bench seat and an adaptable rear compartment. The station wagon, for instance, could carry four additional passengers and 100 pounds whereas the load carrying models will pack 1,000 pounds along with the driver and two passengers up front. On the long wheelbase models the corresponding loads are 10 persons and 400 pounds or driver, two passengers and 2,000 pounds. These pay loads are recommended only when using normal roads. Subtract 200 lbs. and one passenger for rough going.

Power for the Land-Rover is provided by one of two basic units; a 2,250 cc ovh gasoline engine developing 77 bhp at 4,250 rpm with torque of 124 lb/ft at 2,500 rpm, or a 2000 cc diesel of Rover design, a four-cylinder unit developing 51 bhp at 3,500 rpm.

One glance at the Land-Rover chassis is sufficient to gain the impression of ruggedness which is borne out by the vehicle in operation. The chassis is of box section steel with a simple straightforward design of longitudinal and cross members combining to give great strength. Semi-elliptic leaf springs are employed front and rear with the axles passing between the springs and the chassis. The ride is surprisingly good and during the 900-mile trek around the Winter Rally Route we were at no time disturbed by the action of the suspension.

The lowest point of the Land-Rover clears the ground by a good eight inches but because of the alloy structure of the body the weight concentration is low in proportion to the height and the driver quickly learns to overcome the feeling that the vehicle will lean heavily when cornering. In fact the Land-Rover corners well and will cling tenaciously when driven along the side of embankments up to 30 degrees from the horizontal.

The steering is delightfully light, another pleasant surprise the Land-Rover has to offer. Steering is effected by worm and nut with recirculating ball.

The Land-Rover can be driven through either two or all four wheels, the method of engagement being quite simple. The gearbox and transfer box for extra low ratios are situated in the centre of the cab but the position of the long and slack-feeling gear lever favours right hand drive position. When driving from the left hand side there is an appreciable reach into the third gear position which tends to become tiring on long journeys. The main ratios are well chosen but there is more than usual interest for the driver in that only the top two gears are synchromeshed. Finding second gear cleanly does not present too much of a challenge, however. The lower ratio gears can be engaged by moving the red-topped lever to its rear-most position. This must be done with the vehicle stopped and the main gear lever in neutral. Moving back into the main ratios can be done on the move. Front wheel drive is engaged by pressing down the yellow-topped lever next to the gear lever. To return to two-wheel drive the Land-Rover must be stopped and the red-topped lever (low ratio) moved back and then forward.

## LAND ROVER

### SPECIFICATIONS

**Model Tested:** 1961 Land Rover Station Wagon (Regular)
**Price:** $3,195

#### ENGINE

| | |
|---|---|
| Cylinders: | 4, in line, ohv, watercooled |
| Bore & Stroke: | 3.562 x 3.5 |
| Displacement: | 2286 cc, 139.5 cu. ins. |
| Torque: | 1241 lbs/ft. @ 2,500 rpm. |
| B.H.P. | 77 @ 4,250 rpm. |
| Compression Ratio | 7:1 |

#### DIMENSIONS

| | Reguluar | Long |
|---|---|---|
| Wheelbase: | 88 ins. | 109 ins. |
| Track: | 51½ ins. | 51½ ins. |
| Length (overall) | 142½ ins. | 175 ins. |
| Width (overall): | 64 ins. | 64 ins. |
| Height (overall): | 77½ ins. | 81 ins. |
| Turning Circle: | 38 ft. | 45 ft. |
| Ground Clearance: | 8 ft. | 9¾ ft. |

#### CARRYING CAPACITY

**On Road or Rough:** R 7 persons, L 10 persons.
**On Road:** R 3 persons & 1000 lbs, L 3 persons 2,000 lbs.
**In Rough:** R 3 persons & 800 lbs, L 3 persons & 1,800 lbs.

#### DRIVE AND SUSPENSION

**Transmission:** Four forward and reverse, Synchromesh top and third gears only. Transfer gearbox - Two speeds.

| Main Gearbox | Transfer Box | |
|---|---|---|
| | High ratio | Low ratio |
| First Gear | 16.171 | 40.068 |
| Second Gear | 11.026 | 27.742 |
| Third Gear | 7.435 | 18.707 |
| Top Gear: | 5.396 | 13.578 |
| Reverse Gear | 13.745 | 34.585 |

**Power Take-Off:** Centre & rear (extra), Maximum draw-bar pull 4,000 lbs.
**Suspension:** Semi-elliptic plus double-acting telescopic shock absorbers.
**Steering:** Worm & nut with recirculating ball.
**Tires:** Regular - 6.00 x 16; Long - 7.50 x 16.
**Electrical System:** 12 volt, 57 amperes.

---

We found the Land-Rover to be amply geared for highway use and it was found capable of bowling along at more than 70 mph at which speed, however, there was enough vibration to interrupt normal speech. On snow, ice and in mud, the four-wheel drive gave stability and a very satisfying degree of grip.

Both front and rear axles have a ratio of 4.7:1 employing spiral bevel gears.

The driving position is high and affords good visibility. Legroom is not generous but again, drawing from our 900-mile drive in February, was not unduly restrictive. During the Winter Rally tour of duty our Land-Rover's gasoline engine turned in a mileage of 18 mpg, this with sparing use of four-wheel drive.

The test model (station wagon) is priced in Toronto at $3,195 basic with such additional items as spare wheel carrier on hood ($6), water temperature and oil pressure gauge ($32.50) extra windshield wiper and motor ($19.25), hand throttle control, ($4.80) and a fresh air heater ($84). The test model also packed a winch at the front ($489.00) and 150 feet of 5/16 steel cable ($41.15).

Some other optional "extras" which are worth noting when considering the operation of a Land-Rover in winter in this country are the winter insulating trim kits which come at $105.00 for the truck cab models and $85.00 for the hard top.

Among the numerous extras for the Land-Rover is a wide selection of power take-offs.

*On level stretches of the M1 motorway the Redwing appliance touched over 65 m.p.h., 14.5 m.p.g. being returned on a 23-mile full-throttle run.*

OUTBREAKS of fire are never welcome at any time, but less welcome than most is the type of fire that occurs in out-of-the-way spots to which access by large fire appliances intended for use on normal roads becomes virtually impossible. The obstructions likely to bar the way of such vehicles are numerous and include narrow gaps, tight corners, rough or loose surfaces, steep gradients and ice or snow.

A small appliance with drive on all wheels can be valuable under such conditions, particularly if its design is such as to make the most of every available inch of space within quite meagre overall dimensions, whilst at the same time maximum use can be made of a standard vehicle or its components. Such an appliance is the Carmichael Redwing FT/6 forward-control machine based on the Land-Rover 9-ft. 1-in.-wheelbase 4 × 4.

The prototype of this vehicle was originally exhibited at the Fire Officers' Tournament, Conference and Exhibition held in Margate a year ago, and has since been demonstrated to a number of interested bodies by the manufacturers, Carmichael and Sons (Worcester), Ltd., Gregory's Mill Street, Worcester. Production of the Redwing FT/6 is now under way, but it was the prototype that was made available to me for test recently. The standard production model, however, differs very little from its prototype, the main changes concerning only the glazing in the side of the body and slight repositioning of the driver's control pedals.

**Basically Standard Land-Rover**

Being a basically standard Land-Rover beneath its all-enveloping body structure, the FT/6 appliance retains most of the well-known rough-country capabilities of this well-proven design, whilst having a comparatively high power-to-weight ratio for lively performance on metalled roads. Clever adaptation of the standard Rover layout without the introduction of more than a few non-standard components has, however, resulted in an entirely satisfac-

*(Above)* With the low ratio of the auxiliary box engaged, restarts were made in second gear on the 1-in-6.5 section of Bison Hill. The ladder and suction-hose stowage can be seen here.

*(Right)* It took less than a minute to put the first-aid hose into operation. The hose reel is carried above the 140-gal. water tank, and the control panel on the left of the body. Alternatively, this panel can be on the right-hand side.

Carmichael Redwing fire appliance, based on forward-control conversion of long-wheelbase Land-Rover, is compact, manoeuvrable, fast and economical whilst having a good off-the-road performance and ample body space also

# Carmichael Redwing Fire Appliance

*(Below) Despite its fairly high centre of gravity when the first-aid tank is full, the Redwing retains most of the cross-country abilities of the Land-Rover.*

*(Above) High inter-axle articulation is one of the basic requirements of any vehicle expected to operate off the road under severe conditions, and this the Carmichael Redwing certainly has.*

*This view of the engine compartment shows the power unit's position relative to the driving seat, also visible being the gear-change lever and part of its remote-control linkage. Engine access generally is good.*

tory forward-control layout, thereby enabling the Redwing to carry a crew of four (including the driver), a 25-ft. double-extension ladder, two 8- or 10-ft. lengths of suction hose, 12 75-ft. lengths of 2.5-in. delivery hose and 80 lb. of ancillary equipment, such as nozzles, stand pipes and so forth.

Additional to this is a baffled 140-gal. first-aid water tank, above which is mounted a hose reel containing 120 ft. of 0.75-in.-bore hose, while in a recessed control panel on the left side of the body there are two 2.5-in. delivery valves. This panel also contains a hand throttle, priming lever, tank-to-pump valve, compound and pressure gauges, engine oil and water temperature gauges, and a lamp. The panel can be fitted on the right-hand side if required.

There is a choice of three types of pump, the mounting position for any of which is immediately ahead of the rear axle—between the rear seats and the first-aid water tank. The vehicle I tested had a German K.S.B., 350-g.p.m.,

*Filling the first-aid tank is a simple operation, there being a hydrant filler at the rear of the vehicle. A tubular level indicator is provided and this is to the left of the roller shutter.*

## ROAD TEST No. 741/M200—CARMICHAEL REDWING 4 × 4 FIRE APPLIANCE

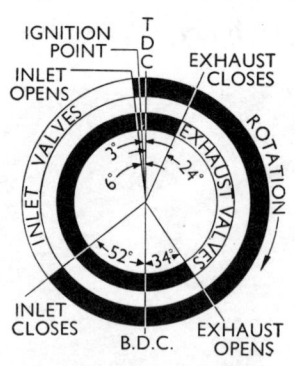

**MODEL:** Carmichael Redwing FT/6 9-ft. 1-in.-wheelbase, forward-control fire appliance, based on Land-Rover petrol-engined 1-ton 4 × 4 chassis.

**WEIGHTS:**

|  | Tons | cwt. | qr. |
|---|---|---|---|
| Unladen (kerb weight) | 2 | 0 | 0 |
| Equipment, etc. |  | 19 | 2 |
| Driver, observer, etc. |  | 3 | 2 |
|  | 3 | 3 | 0 |

**DISTRIBUTION:**

| | Tons | cwt. | qr. |
|---|---|---|---|
| Front axle | 1 | 5 | 1 |
| Rear axle | 1 | 17 | 3 |

**ENGINE:** Rover four-cylinder o.h.v. petrol engine; bore 90.49 mm. (3.562 in.); stroke 88.9 mm. (3.5 in.); piston-swept volume 2.286 litres (139.5 cu. in.); maximum net output 77 b.h.p. at 4,250 r.p.m.; R.A.C. rating 20.5 h.p.; maximum net torque 124 lb.-ft. at 2,500 r.p.m.

**TRANSMISSION:** Through 9.0-in.-diameter single-dry-plate clutch to four-speed synchromesh main gearbox and two-speed transfer box, thence by one-piece propeller shafts to the fully floating spiral-bevel front and rear axles.

**GEAR RATIOS:** Main box; 2.996, 2.043, 1.377 and 1 to 1 forward; reverse 2.547 to 1; transfer box; 2.888 and 1.148 to 1; axle ratio 4.7 to 1.

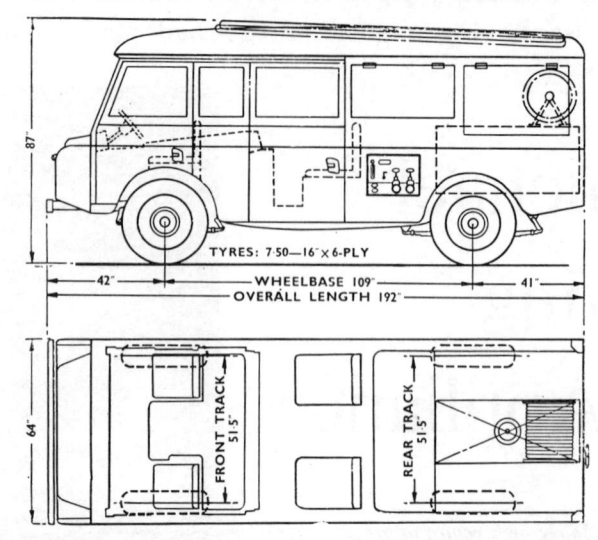

**BRAKES:** Girling hydraulic system, with leading-and-trailing-shoe units at all wheels. Single-pull handbrake linked mechanically to 9.0 in. × 1.75 in. drum brake on rear of transfer box. Diameter of drums, front, 11.0 in., rear, 11.0 in.; width of linings, front, 2.25 in., rear, 2.25 in.; total frictional area, 172.0 sq. in., that is, 54.5 sq. in. per ton gross weight as tested.

**FRAME:** Welded box section, with eight cross-members welded in position.

**STEERING:** Burman recirculating ball; 4 turns from lock to lock.

**SUSPENSION:** Semi-elliptic springs, with telescopic dampers at both axles.

**ELECTRICAL:** 12 v. compensated-voltage control system with 57-amp.-hr. battery.

**FUEL CONSUMPTION:** (a) normal speed, 20.0 m.p.g. at 29.5 m.p.h. average speed; (b) motorway, full throttle, 14.5 m.p.g. at 54.5 m.p.h. average speed, that is 63 gross ton-m.p.g. as tested (a) and 45.7 gross ton-m.p.g. as tested (b), giving time-load-mileage factors of 1,859 (a) and 2,491 (b).

**TANK CAPACITY:** 10 gal., normal-speed range approximately 200 miles.

**ACCELERATION:** Through gears, 0–20 m.p.h. 6.25 sec.; 0–30 m.p.h., 13.25 sec.; 0–40 m.p.h., 22.5 sec.; top gear, 10–20 m.p.h., 18.0 sec.; 10–30 m.p.h., 33.0 sec.; 10–40 m.p.h., 49.0 sec.

**BRAKING:** From 20 m.p.h., 16.0 ft. (27.1 ft. per sec. per sec.); from 30 m.p.h., 40.0 ft. (24.2 ft. per sec. per sec.). Handbrake from 20 m.p.h., 29 per cent. (Tapley meter).

**WEIGHT RATIO:** 1.22 b.h.p. per cwt. gross weight as tested.

**FORWARD VISIBILITY:** To within 9.5 ft. of front bumper at ground level on centre line.

**TURNING CIRCLES:** 46.5 ft. left lock, 48.5 ft. right lock. Swept circles: 49.5 ft. left lock, 51.5 ft. right lock.

**MAKERS:** Appliance and forward-control conversion by Carmichael and Sons (Worcester), Ltd., Gregory's Mill Street, Worcester: basic chassis by The Rover Co., Ltd., Solihull, Warwickshire.

---

two-stage centrifugal pump with built-in water-ring primer. Similar output is given by the second alternative, which is a Coventry Climax F.W.P. 300/350-g.p.m. centrifugal unit with automatic water-ring primer. The third option is also of Coventry Climax manufacture and is a U.F.P. 500-g.p.m. pump. In all cases the pump is shaft-driven from the standard Land-Rover centre power take-off and the pump-engagement lever is mounted on the cowling to the left of the driving position.

For the most part the modifications to the standard Land-Rover consist of a welded box-section forward frame extension which carries the steering gear and pedal controls, so that the driving seat can be positioned immediately above the front wheels. The standard steering layout has remained virtually unaltered except for the use of a rearward-facing draglink.

On the prototype tested the closeness of the three pedals was found to be a disadvantage, but production models will have a slightly different arrangement which permits the pedals to be more widely spaced.

The engine and radiator location are the same as on the standard vehicle and the remote-control gear-lever linkage is attached to the engine so that the knob lies approximately above the radiator. The engine-radiator compartment is enclosed by a rear-hinged T-shaped glass-fibre cowl, which can be removed completely. Ahead of this cowling a box-section duct continues forward to the radiator grille at the front of the vehicle. The top panel of this duct is joined to the lower edge of the facia panel and incorporates the auxiliary gearbox change-speed lever, the front-wheel-drive engagement knob and the choke control, in addition to the pump-engagement lever.

Ahead of the flat one-piece windscreen there is a small hinged " bonnet " which gives access to the hydraulic-fluid reservoir, the electrical control box and fuses, and the brake and clutch master cylinders. The scuttle assembly retains the usual Land-Rover hinged vent panels beneath the windscreen and incorporates a glove pocket on each side of the central instrument panel.

The driving seat is mounted on longitudinal slides which afford a good range of adjustment, and access to and from the seat is good. The forward-hinged door opens through a useful angle.

Behind the two front doors there are louvres to prevent heat build-up inside the engine compartment, and to the rear of these there are two standard Land-Rover forward-hinged doors which give access to the rear crew compartment. Although there is room in this compartment for three seats abreast, there is a trap between the two seats, and this gives access to the water pump.

Behind the rear-compartment doors on each side of the vehicle there are two drop-down, hinged flaps, and these give access to the equipment-storage space alongside the first-aid tank. At the rear of the body there is a fixed tailboard across the lower half, the upper section being enclosed by a roller shutter.

With the exception of the engine-cowl top and the roof, all the bodywork is of aluminium. The roof is a one-piece glass-fibre moulding with a full-length trough, in which the extension ladder and the suction hoses can be stowed. Immediately above the engine compartment this trough incorporates a detachable aluminium panel to facilitate engine removal, whilst ahead of the trough twin amber flashing lights are fitted as standard. Provision is also made for mounting a siren or bell.

Intended to operate at a fully equipped gross weight of approximately 3 tons, the Redwing FT/6 has 7.50-16 (6-ply) semi-cross-country tyres and Rover heavy-duty rear springs. These springs have a rate of 360 lb./in. and are offered as optional equipment on the standard long-wheelbase Land-Rover. The front springs are of special design, however, and have the same rate as the rears. Suitable dampers are provided in each case.

In the version tested the Carmichael FT/6 machine is suitable for use as a second-line appliance by local fire brigades (Cheshire Fire Brigade have recently ordered four) and also by industrial and forestry undertakings. An airfield crash-tender version can be supplied with dry-powder extinguishing equipment replacing the first-aid tank.

Stripped of all equipment the FT/6 appliance has a kerb weight of 2 tons, of which the front wheels carry 1 ton 1 cwt. For my test the extension ladder and suction hose were in position on the roof and the first-aid tank was filled with water, but the weight of other items of equipment and that of two crew members in the rear compartment was represented by sand-filled wooden cases which were evenly distributed about the rear of the vehicle. This imposed weight increased the rear-axle loading by 18.75 cwt., but added only 0.75 cwt. to the front-wheel loading, this latter weight being increased to 1 ton 5.25 cwt. with myself, a Carmichael representative and test gear.

The front/rear weight ratio was not ideal, therefore, but this has already been reckoned with and production versions of the Redwing have the first-aid tank and hose about 6 in. farther forward than on the test vehicle to increase the weight on the front axle.

From the performance figures obtained on normal roads (detailed in the accompanying data panel) it will be seen that the appliance offers a combination of liveliness, economy and safety. Despite the considerable weight of water in the first-aid tank and the rather high mounting of this tank, the machine was quite stable and there were no undesirable side effects when making full-pressure braking stops from 20 and 30 m.p.h.

The road-performance figures obtained are good and compare favourably with the results obtained last September with a standard short-wheelbase Land-Rover. Naturally enough, the acceleration of the appliance was down on that recorded with the smaller vehicle, but there was a difference in weight of 1 ton 2 cwt., despite which the Carmichael vehicle stopped from 30 m.p.h. in a distance only 7.5 ft. greater than that of the lighter, standard machine. Approximate gear-speed checks suggested that 25 m.p.h. could be reached in bottom gear, 36 m.p.h. in second, 52 m.p.h. in third and over 65 m.p.h. in top, these latter figures being timed while conducting the 23-mile full-throttle consumption test on the M1 motorway.

Gradient performances were verified on the 0.75-mile Bison Hill, the average grade of which is 1 in 10.5. A maximum-power ascent in an ambient temperature of 14° C. (57° F.) was made in 2 minutes 22 seconds, bottom gear being used for 38 seconds, during which time the road speed did not drop below 18 m.p.h. The effectiveness of the cooling system was shown by the fact that the engine-coolant temperature was the same at the end of the climb as it had been at the beginning—70° C. (158° F.).

*Good driving conditions are given by the Carmichael forward-control conversion of the standard Land-Rover long-wheelbase chassis, and access to the driving seat is easy. The cowl passing beneath the facia panel forms an air duct for the radiator.*

Fade resistance was checked by coasting down the hill in neutral while keeping the footbrake applied to restrict the speed to 20 m.p.h. This descent lasted 2 minutes 40 seconds, and at the bottom of the hill a stop from 20 m.p.h. produced a Tapley-meter maximum-deceleration reading of 86 per cent., indicating that virtually no fade at all had occurred when this figure is compared with the average reading of 87.5 per cent. obtained earlier when the drums were cold.

Gradient restart tests were made on the steepest section of Bison Hill, the gradient of which is 1 in 6.5. Facing up the hill, the handbrake (which had not been adjusted recently) was not quite able to stop the vehicle from rolling slowly backwards. The highest gear-ratio combination in which a restart could be made was found to be second, with the low auxiliary-box ratio engaged. Attempted restarts in second-high and third-low resulted in clutch slip. Facing down the hill, the handbrake proved powerful enough and several reverse-low restarts were carried out. Because of the ratio multiplication provided by the two gearboxes, restarts should be possible with the Redwing on gradients approaching 1-in-2 severity, using the two lowest ratios available.

Taken off the road, the Redwing was driven over some quite rough ground without faltering and good rough-ground traction characteristics were displayed. The first-aid hose was then put into operation, less than a minute elapsing between the time that the procedure was commenced and the time that the hose was in use. To keep the engine cool during long periods of pumping with the vehicle stationary, an engine-lubricant heat exchanger is incorporated in the pump circuit, and after five minutes with the first-aid hose in use the engine oil and water temperatures were still normal.

On the road the Redwing handles well and is safe and comfortable to drive. Despite the slab-sided bodywork and the relatively high centre of gravity the appliance is stable at high speeds and the steering is precise but light. Engine noise and heat levels are entirely acceptable and access to the power unit quite reasonable.   J.F.M.

*Worm's eye view of the Roadless 109, a sturdy and most impressive vehicle for the worst rally routes.*

## ROADLESS IMPRESSION

# IDEAL RALLY CAR?

FOLLOWING a stimulating argument with John Sprinzel about the ideal rally car, and with Jack Kemsley already plotting more routes over Forestry Commission property for the 1962 RAC Rally, we decided to tackle the problem ourselves.

This was to be no dilettantic toying with the business; we would do it seriously.

After an intensive top-level conference, during which scores of specifications were studied and two of the staff complained of migraine, we were suddenly given a lead by Gethin Bradley, the genial young man who copes with the problems which journalists take to the Rover Company at Solihull.

The result was that one bitingly cold day this month we found ourselves on a windswept Hounslow Heath, all keyed up for a test worse than any of the sections on this year's RAC Rally.

### TANK DESIGN

It had not, after all, been necessary to design and build our own rally special. That had been done for us by Roadless Traction Limited, of Hounslow. And when I tell you that the company's chairman is Colonel Phillip Johnson, who played a big part in tank design during the first World War, you will readily appreciate that we had a machine equipped to take punishment.

John Sprinzel feels some misgivings about the rally potential of the BMC minis, with their ten-inch wheels. To meet that criticism, our test machine had gone up a size or two, using 10.00 by 28 tractor tyres, and giving a useful ground clearance of 17 inches under the differentials.

Front-wheel drive or rear drive? Oh, the arguments we had about that! As an answer to these conflicting opinions the Roadless 109 has—like the Land-Rover on which it is based—four-wheel drive, with the customary option of rear drive for fast motoring on good roads.

In order to keep the navigator awake, the Roadless 109 has eight forward ratios, so that the driver can continually be asked if he is in the right gear. There is a further subtle feature that should do away with any need for wakey-wakey pills—the seats are bolt upright, and there is no possible danger of the map-reader nodding over his labours.

It was immediately obvious that the Roadless 109 met all our demands on the score of toughness. Perhaps it **is** a trifle heavy by today's rally standards—tipping the scale at close on two tons—but there should be no need for a horde of unbenders and panel bashers to station themselves at strategic points along the route. And with an overall width of 7 feet 6 inches, woe betide anyone daring to attack any section in the wrong direction.

Neither should there be any difficulty in a situation where you round a bend and find nine competitors completely blocking the road. There are two alternatives; either you drive right over them, or take to the country.

This struck us as one of the prime virtues of the machine, for it proved virtually unstoppable in conditions in which the normal cissy rally car would bog down and probably disappear from sight. The Roadless 109 proved capable of wading through water up to 2 feet 6 inches deep, while it growled contentedly through bogs and soft sand.

With the driver's head some six feet from ground level, visibility was extremely good. Coupled with an immense chassis frame and sturdy front bumper, it engendered a feeling of complete confidence. The instruments were well placed, though perhaps the pedals were a trifle large and the steering rather low-geared for driving tests. The turning circle, too, might prove a little embarrassing in tight corners—it is around 40 feet—but after all, what is a tree or two when one is rallying in earnest?

On some of the really tight sections the machine's maximum speed might be considered a drawback. The Rover Diesel unit fitted to our test vehicle was tending to scream its head off when we were rolling along in top gear (13.29 to 1) at 28 miles an hour. But the Roadless 109 will cruise all day on normal roads at around 23 miles an hour, and at that speed M1 can be traversed from end to end in about three hours.

But it is on surfaces rougher than this that the machine comes into its own. There would be no complaints of any of the Forestry Commission sections if everyone competed in one of these. (In fact, the Forestry Commission have one already, which leads several clever people to suggest that Jack Kemsley used it for his reconnaissance of this year's route).

Plunging through a mere foot of water, I set the Roadless 109 at a 45-degree bank of clay and gravel. For this rather simple climb I selected second gear in the lower range—seventh, if you like. There was absolutely no need to rush the slope; we mounted remorselessly at an estimated two miles an hour, seeing nothing but sky from the driver's seat. Carlsson would have been most impressed, I'm sure, but he would have some difficulty in putting the Roadless 109 on its roof. I tackled a 1 in 1 downhill slope at an angle, encouraged by Mr. Rew, who is Colonel Johnson's personal assistant, and was surprised, (and relieved), to find myself safely at the bottom the right way up.

Under these sort of conditions, the brakes rarely need dabbing. But consideration is being given to fitting discs, since the drums tend to fill with water and mud when the vehicle is gambolling in its natural habitat.

Tyre pressures are somewhat critical. For our roadless rallying we had them at around six pounds per square inch. When used on normal roads the machine handles more satisfactorily with the pressures set at 12 or 13 pounds psi.

There is at present no provision for a spare wheel. But need this be a worry? Mr. Rew assured me that up to date they have not had a single puncture, and even if one smacks a kerb or boulder it is not the tyre which suffers!

The suspension is by no means soft on its 'cart' springs, and the absence of sound damping might be criticised by some of the mollycoddled rally boys. With the multiplicity of gears it is sometimes difficult to judge the engine speed, but the firm is considering fitting a rev counter.

### FUEL CONSUMPTION

It might be necessary to fit an auxiliary fuel tank, since the Roadless 109 covers about ten miles to each gallon on long runs, and less when mud plugging. The tank, by the way, holds 15 gallons.

The machine is also available with the Land-Rover petrol engine, which gives snappier performance. Plans are in hand for fitting a higher ratio to give a top speed of over 40 miles an hour in the petrol version.

During some fairly lively motoring over a smooth surface it was possible to assess the handling. Throwing the machine into a corner at 15 miles an hour revealed definite understeer. At low speeds the steering was a trifle heavy, and power steering might be an advantage for wiggle-woggle tests. The brakes were adequate for all demands I made on them.

The Roadless 109, costing £1,548 for the petrol version, and £1,648 for the Diesel, (chassis and cab only), is obviously a vehicle with a great potential for the intrepid and adventurous.

Providing the regulations can be stretched, this is a most desirable rally machine. It may not be a winner in any concours d'elegance, but it should always be a finisher.

The BBC, I am told, have one on test at the moment. So that's what Raymond Baxter's up to . . . .

A.B.

*Story by RAB Cook   Pictures by Maurice Rowe*

# Horses for courses

## 'Motor' goes testing in the rough

A THREE-DAY testing session, snow-covered Perthshire hills and one of the oddest-looking collections of vehicles ever assembled have proved two things: (1) there are on the British market vehicles suitable for practically any kind of surface you are likely to encounter in this country; (2) no single vehicle can cope with *all* types of going. As the title of this story indicates, it is very much a matter of horses for courses.

The horses in this case were the Austrian built Haflinger—truly a "horse" for it is named after a breed of pony; the Swedish Snow-Trac; Land Rover, in both short- and long-wheelbase forms; short-wheelbase Austin Gipsy; Renault 4L Estate; and the B.M.C. Mini-Moke. The object of the operation was not only to discover what each of these could do, but to winkle out the things they couldn't—or shouldn't be asked to try. Individual reports come later but, first, the story of a unique three days.

The vehicles, with drivers supplied by their various manufacturers and concessionnaires, descended like a plague of locusts upon the 8,000-acre Ochtertyre Estate, near Crieff, owned by Sir William Keith Murray. Despite the fact that we cut grooves all over his mountains, completely ruined a ski run and put the fear of death into hundreds of sheep, grouse and mountain hares, Sir William not only took things in good part but joined in with great gusto, using his own Snow Trac. I asked him if he thought that over £2,000 was dear for a Snow Trac. He replied: "Well, it was that or £28,000 to build a road."

The days were utilized thus: on the first, we set off in the general direction of "up" to seek out really nasty hazards on a circular route; the second day, we aimed ourselves at a 2,000 ft. heather-clad hill, trying to get there by side-stepping the hazards whenever possible; day three was when everyone tried everyone elses' vehicles with considerable elan on upland sheep-grazing. From the air, that meadow must have looked like a withered lettuce leaf alive with multi-coloured fleas.

Off, as the saying goes, we set, with Snow Tracs front and rear. I went, initially, as passenger in the Renault out of sheer morbid curiosity to see how far it would get.

Soon we came alongside a hummock with a two-in-one slope, grass with frozen earth below, and a surface far from smooth. At the bottom of it was snow, about ten inches deep. The s.w.b. Land-Rover detached itself from the convoy and made a rush at this hill and would have clawed up easily but for one thing—the crisp snow was as good a speed killer as dry sand. So nearly there . . . . The Gipsy driver wasn't having this but he got out first and inspected the hill before charging it and just making the summit. Several more tries and the Land-Rover did the same and all the while the two Haflingers were making apparently futile attempts. Then they got the idea of letting their tyres down to 16 p.s.i. and the hill was again conquered. Honours even.

Meanwhile, in another place: both the Renault and the Moke had spun their front wheels to a stop on the main route and only diagonal progress upwards was possible. I had a sudden idea, sitting in the Renault: "Try it backwards!"

**Where General Wade declined to go . . . Land-Rover and Gipsy high in the hills of Ochtertyre.**

# Horses for courses

"Why not?" and we were off. The Mini-Moke driver got the idea and soon the two of them were flashing up the track in reverse with far more traction than they could get the right way round. Later we heard that a local farmer does just this trick with his own 4L when delivering food supplies to upland sheep.

On. Now in the Gipsy, I saw, just over the brow of a hill, that the s.w.b. Land-Rover had dug its wheels into deep mud and that the driver was busily driving steel stakes through a flat piece of perforated steel, and down into the ground. This was a ground anchor to which the winch cable was attached and, despite the soft peat, the vehicle winched itself out at the second attempt—insufficient stakes foiled the first try.

So I transferred to this Land-Rover and we followed the Snow Trac. But why were the people ahead grinning at us so widely all of a sudden? We grinned back and waved—then it happened. The Snow Trac glided on, leaving hardly a mark on the ground, but we dived nose first into a bog with one front wing nearly out of sight, and sinking slowly. It took two Snow Tracs and one Land-Rover, coupled together, to haul us backwards out of that one. Moral: don't trust (a) virgin snow and (b) grinning Snow Trac drivers.

We went no farther in that direction. A ridge some three feet high provided an interesting test. Here the l.w.b. Land-Rover buckled the valances under the doors but the Haflingers, with their short wheelbases, didn't really regard it as a ridge anyhow, and nipped over. The Gipsy and s.w.b. Land-Rover got over easily and the others had more sense than to try.

On the return route we got the first sight of a little ceremony peculiar to the Haflingers, which was to be repeated many times. They would scramble down into a stream bed and then shunt back and forth a few times. Stuck? Not on your life—washing themselves like sparrows in a bird bath!

We had learned this much: Land-Rover, Gipsy and Haflinger will get through mud of considerable depth if there is a hard base somewhere below but bottomless peat bogs are Snow Trac country. Deep snow is a monumental speed killer, especially if it is over heather, and it will eventually bring any wheeled vehicle to a halt. If you do stick in soft snow or reasonable mud, immediately go backwards—further attempts at forward motion mean bogging in but reversing and charging a few times will usually get you through. On steep places with good grip, the trick is to go slowly and thoughtfully rather than charge like a maniac—indeed, charging should always be a last resort. All the vehicles will cope with boulders.

The Renault got farther than I had thought its engine power would take it and the Mini-Moke didn't suffer much from its small ground clearance; tractor-type tyres would have enabled it to accomplish more.

The second day brought the assault upon the unsuspecting 2,000 ft. mountain. On the way up, things seemed to be going too well. When the hard, rutted track gave way to slippery heather, we abandoned the Renault and the Moke—no sense in damaging them—and noted that, in the meantime, the Snow Tracs were contemptuously ignoring the track and padding on over the snow-clad heather.

We came to a little glen and, with a certain degree of cold fear, I learned the wheeled method of taking a down-hill, snow-clad heather slope. You charge it! Braking (using a low gear) is just like that on the level but I took a lot of convincing. Several uphill slopes which I thought certain to defeat the wheeled vehicles were conquered—not always at the first go.

Then, with the summit in sight, we came to It: a slight gully, filled with snow, which had to be crossed with the nearside wheels quite a bit higher than the offside ones—and

*Will he, won't he? The Mini-Moke charges a burn with mud and snow at the exit side. It kept going in great style.*

*Same spot as above with the Renault taking the plunge. It upheld front-wheel-drive honour and shook itself clear like a bedraggled spaniel.*

"I launched a Snow Trac in the air..." In such circumstances, this astonishing vehicle falls to earth in a sedate and gentle manner—the expected crunch never comes.

there was little or no run in. The danger was of slipping or tipping over to the right.

The s.w.b. Land-Rover had a go, slithered, stopped for safety reasons and was winched through by hitching the cable to a Snow Trac already on the other side. Then, disaster... the Gipsy and the other Land-Rover were both stuck side-by-side in the snow, about eight inches apart, and the Land-Rover had the tinned soup and cooking stove aboard and had to be on the level to go into culinary action. And it was lunch time! Both were hurriedly winched out backwards, soup served and the Gipsy made a most spectacular charge through the now well-broken snow, almost tipped up, but made it.

Meanwhile, the first Land-Rover had sneaked off round the back of the hill and suddenly appeared triumphantly on the summit. Obviously, it was going to get crowded up there! The Haflingers started charging the snow; then they retired to fit chains and succeeded, again looking as though a roll-over was imminent. So we had them all through except the l.w.b. Land-Rover which we didn't want to risk in view of the rations aboard. It was really bleak up there.

Alongside the summit was a 45-degree slippery rock face which defeated the Snow Trac—the only thing that did—but it went round another way and came down it. This was too much for one Haflinger which followed suit, using the cut-away nose as a skid for the levelling out at the bottom.

Back down through a flurry of snow, to find that part of the route had deteriorated into an apparently bottomless bog and both Land-Rovers stuck in it—the s.w.b. one after getting through and then returning to rescue its mate. The others had been lucky in finding alternative paths but it's quite a thing to watch a Land-Rover slowly sinking into a Highland marsh. They tell me they've lost whole horses that way.

Lessons? Mainly that if you're going to try the really way-out rugged stuff, you must take two vehicles and enough savvy to prevent them both being bogged down at the same time. Or you can take a single Snow Trac, but then you have to transport it to the roughery on a trailer.

The findings of the final day follow in the form of short individual test reports. It says a lot for the cross-country vehicle of today that although I had never driven one before, car experience was quite enough to give immediate command. In another few years they'll have automatic transmissions, GT versions, and, let's hope, echo-sounders to detect hidden peat bogs.

It would be churlish to close this part of the story without mentioning the Singer Chamois which photographer Maurice Rowe and I used for generally nipping around Perthshire. The amount of sheer grip available on frozen grassland hills astonished everyone and, if it did stick, it was only through lack of low speed engine torque or the smell of hot clutch. On the level it could whip itself off glazed ice or out of sticky mud very smartly—and it was on standard road tyres. I'd love to try one with cross-country tyres and lower gearing.

# Horses for courses *Continued*

## SNOW-TRAC

**Concessionaire:** Innes Ireland, Downton House, Walton, Presteigne, Radnorshire.
**Agent:** Alexander MacLarty, Contractor, Galvelmore Street, Creiff, Perthshire, Scotland.
**Specification:** 1,200 c.c. flat-four, air-cooled o.h.v. Volkswagen engine; four-speed and reverse gearbox with synchromesh on all gears.
**Weight:** 21 cwt.
**Basic price:** £2,047 7s. 3d. (inc. p.t.). Total with cabin, seats, second screen wiper, hour-meter, etc. as tested, £2,160 7s. 3d.

DESPITE its somewhat Bren Gun Carrier appearance, the Snow-Trac is relatively light at 21 cwt. This is achieved by things like an aluminium-clad body and tracks which are in effect twin fabric-reinforced rubber bands with metal cross-pieces which take the drive via the forward cogged wheels. The other 14 wheels are rubber shod.

The result of all this ingenuity is a ground loading of merely ¾-lb. per square inch so that even on soft snow it leaves impressions barely one inch deep and it treats water-logged peat in like manner.

Sitting high behind the front-mounted engine the driver has a splendid view and can see the ground very close to the nose—an important point. Its progress is relentless: the flat-out speed from the governed engine is 15 m.p.h. but in the right conditions—such as charging down a heather-clad 1-in-2 slope—this can be as exciting as 120 m.p.h. on the M1.

It is natural to approach this device with some trepidation but the conventional steering wheel and pedals (on the left) are reassuring. The four-speed gearchange is inverted—i.e. with first at bottom right—but this is the only thing which requires extra thought. The steering proved to be feather light if slow to respond and 10 yards' progress was enough to make me feel completely at ease. Variation of the gearing to the individual track drives is the method of steering.

The brakes, I was warned, weren't much good—which proved to be true unless one really stood upon them hard—but it didn't matter as a quick switch into the synchromesh first gear provided all the retardation necessary. General noise level was fairly high, but not bad enough to make conversation difficult. The more people carried, the quieter it seemed; the total load is seven, including the driver, but the same again can be towed on a sledge and a great deal of spare equipment stowed around the outside.

Fuel consumption is one gallon per hour and the tank capacity is seven gallons. The only conditions likely to defeat it are hard ice or a rock face around the 45 degree mark; and a marshy or snow-filled hole under one track only could bog it down. The driver must keep a sharp lookout for large boulders, as these would damage the tracks. Track life is said to be four years of average use.

The heater is adequate but little more with only two aboard but the surprisingly supple suspension gives far more comfort than might be expected. Quite unique, and very impressive.

## Test Conditions

All the vehicles were first tried over a upland sheep-grazing meadow which contained frozen, rutted tracks, large hummocks, minor ditches and snow-filled hollows. Then, with the exception of the Snow-Trac, they were driven over a circuit which included an unsealed road; smooth tarred surfaces—one section being frost covered—with moderate hills; and tight turns in wet mud of various depths.

*The inevitability of gradualness—the Snow-Trac just creeps on and on, whether up hill or down.*

---

## AUSTIN GIPSY

**Manufacturers:** The Austin Motor Co. Ltd., Longbridge, Birmingham.
**Specification:** 2,199 c.c. four-cylinder in-line water-cooled engine. Four-speed gearbox with synchromesh on three upper ratios plus two-speed transfer box, giving eight ratios in all. Two- or four-wheel drive.
**Weight:** 24 cwt. (hard top).
**Alternative versions:** Long wheelbase; soft-top, pick-up etc., totalling 38 variations in all. Optional 2,178 c.c. diesel engine.
**Prices:** £680 (no p.t.) soft top. Version tested, hard top, £730 (no p.t.). Extras include power take-off, winches, tachometer, fire pump, etc.

THE original Austin Gipsy couldn't be described as a riotous success but the current Mk. IV model, with steel leaf springs and the fruits of steady development—the G4 M10—is a splendid vehicle. The body is of steel, on an immensely strong chassis, and the unusually close-to-centre-line mounting of the springs permits a very high degree of angular axle movement, clearly evident when clambering over boulders or crossing ditches at an angle. The hard top is of glass fibre.

First impressions were of a comfortable driving position, light controls, positive steering and, for the type of vehicle, fairly lavish original equipment including such items as ash trays, trip mileometer, passenger grab handle, temperature gauge, etc. The effective heater is of the fresh-air type and the passenger is safe from sharp edges and cold metal. In fine weather, the complete top, windscreen and door tops can be removed. Entry and exit are easy. The cab was outstandingly draught proof.

On the rough, the suspension seemed slightly harsh and there is little doubt that a 5 cwt. load in the back would have made a vast difference. It must be emphasised that this was an almost new vehicle—the springs would settle a little with use—and that it was being driven fast in extreme conditions.

On any type of road the steering was light and positive and near saloon car standards despite the cross-country tyres. The brakes were well up to the task and the pedal pressure light: lightest of all was the accelerator pedal and I think that if I owned a Gipsy I would even load the pedal a little with an extra spring.

At 60 m.p.h. the Gipsy sang along happily, with a fair amount of engine noise and some vibration detectable through the steering wheel—it came from the tyres. Changes from two- to four-wheel drive can be made on the move and, by doing this, any tail skittishness on icy corners was brought under control and dry-road speeds could be maintained. An over-all 20 m.p.g. would be near the mark.

In the lower ratios provided by the transfer box, standing starts on greasy slopes were no more difficult than on the level, thanks largely to the good low-speed engine torque and the sensitive clutch.

The thing I didn't like was no fault of Austin—British purchase tax regulations preclude the fitting of side windows in the hard top so that the view to three-quarters rear left, was nil.

The general feeling was one of confidence in a solid vehicle which, even in extreme circumstances, would do what it was told.

*Close coupling of the springs enables the Gipsy to achieve some startling wheel angles on the rough, yet remain roll-free on the road.*

50

## LAND-ROVER

**Manufacturers:** The Rover Co. Ltd., Solihull, Warwickshire.
**Specification:** 2,286 c.c. four-cylinder, in-line, water-cooled engine. Four-speed gearbox with synchromesh on three upper ratios plus two-speed transfer box, giving eight ratios in all. Two- or four-wheel drive.
**Weights:** Short wheelbase estate, 29¾ cwt. Long wheelbase with canvas hood, 29½ cwt.
**Alternative versions:** Pick-up, hard top, estate, fire-pump, etc., available in both long and short wheelbase. 12-seat personnel carrier on long wheelbase free of purchase tax. Optional 2,286 c.c. diesel engine.
**Prices:** 88 Station wagon, £1,010 10s. (inc. p.t.); 109 Long Wheelbase, £790 (no p.t.); s.w.b. soft top, £691 (no p.t.). Extras include winches, pumps, compressors, snowploughs, etc.

ALTHOUGH the Land-Rover has a steel chassis and scuttle, the majority of the body parts are of aluminium, a fact which does not seem to be generally known.

Climbing aboard the short-wheelbase version, I got an immediate feeling of being fully in command of the situation. Under way, the engine and gear noise level was surprisingly low. I would have liked the steering wheel about two inches farther forward as the somewhat low-geared steering ratio calls for a big fistful of wheel when a boulder is suddenly spotted in the way ahead, or during skid correction. On the l.w.b. version, the seats were adjustable and set just right for me.

In both, the clutches initially felt on the heavy side for my shoe-clad feet but, undoubtedly, a pair of hefty boots would make things perfectly O.K. The seats were comfortable and the recirculating heater produced a good blast of hot air, especially comforting to the front passenger. I think that on one occasion, it saved my life!

On the rough, normal bumps were absorbed well by the springs, and the dampers permitted no bounce. The long version was, naturally, more comfortable at speed on the bumps and I didn't feel perched so high in it—lack of the nose-mounted spare wheel may have been the reason. The clutch had ample sensitivity and starts free from wheelspin or judder were simple on slime or deep mud. Using four-wheel drive made no detectable difference to the steering effort on grassland.

The gearchange was quick and positive—a bit of a clank on the change into second—and the controls for the four-wheel drive and transfer box were within easy reach.

On the road: the steering of the short model felt much more positive on the unsealed road and on ice when four-wheel drive was engaged. With the l.w.b., it didn't seem to make much difference and, in both, the low gearing was again apparent when making steering corrections. Nevertheless, flat-out driving on main roads, despite the "hairy" tyres, was a very relaxed business—and comfortable. Fuel consumption—about 18–20 m.p.g.

The one dull note was the siting of the handbrake—it is to the left of the driver, protruding from beneath the seat, which is perfectly all right, but it is too short so that hill take-offs call for bending forward in a nose-to-steering-wheel position to release it.

Otherwise, I could find nothing which didn't fit in with the Land-Rover's world-wide reputation for toughness and its ability to go, or winch itself out of, practically anywhere.

*A sight familiar the world over—a Land-Rover clawing its way over the scenery. This is the long-wheelbase version.*

*In such conditions, the Haflinger overcomes lack of engine capacity by crawling with both front and rear differentials locked. The slope is roughly one in two.*

## HAFLINGER

**Concessionaires:** Ryders Autoservice (G.B.) Ltd., Knowsley Road, Bootle, Liverpool, 20.
**Specification:** 643 c.c. flat-twin, air-cooled o.h.v. engine, mounted behind rear axle. Four-speed all-synchromesh gearbox with additional "crawler" gear optional. Two- or four-wheel drive with differential locks.
**Weight:** Approximately 14¾ cwt.
**Basic price:** £678 (no p.t.). Extras include canvas cab, glass-fibre cab, body sides, rear seats, power take-off, winches, heater, etc.

JUST how a little lorry driven by a 643 c.c. flat twin can clamber over piles of boulders, come down 1-in-1 hills, climb out of stream beds and generally behave like a fly in a bowl of lump sugar is explained when its design is studied.

It has something none of the others has—two pull-up controls which lock both front and rear differentials so that it has *true* four-wheel drive. Normal two-wheel drive is used on the flat. But that is only the start: it has high ground clearance with sharp cutaway front and rear so there is no fear of digging in at the end of a descent.

The chassis is a single fore-and-aft tube from which sprout the swinging arms which carry the wheels above their axle centres. Soft coil springs with rubber buffers inside them plus telescopic hydraulic dampers keep the suspension under control while permitting extraordinary wheel angles.

Finally, the option crawler gear, giving a top speed of 2 m.p.h., allows it to climb anything on which the wheels will grip. Top speed is around 40 m.p.h. (the engine is governed) and about 25 m.p.g. could be expected over mixed going.

The Haflinger needs a fairly long "getting to know you" session. One feels perched rather high and the control pedals have had their positions dictated by the vehicle's shape rather than by ergonomics so, to start with, quite a bit of fumbling is required to find the brake and the same applies to the trip back to the accelerator.

On all types of surface the steering is light and, on the rough, four-wheel or "all-four" drive makes little difference to it. Incidentally, one can select locked rear differential and unlocked front, or two-wheel rear drive with locked diff.!

Crawling out of frozen ruts was easy and the length of leap off hillocks was governed only by the speed available. On the road, the brakes were rather lacking in power but a quick change down to second with the positive lever produced all the retardation one would ever be likely to need—the gap between second and third is considerable. The engine governor was rather a pest when trying to find the best ratio on a main road hill—top was too high and just when a nice speed was nearly reached in third, the power was restricted.

The most unexpected feature was the way in which this mechanical pony scampered round corners on the road and went precisely where it was pointed. An attempted skid turn through 180 degrees on three inches of mud resulted in no skid, no tipping over on the side, but a mighty quick turn!

It would take me a week to get used to a Haflinger so that its full capabilities could be revealed. But the things other people did with it impressed me immensely and clearly emphasised the benefits of combining lightness with four-wheel traction, a smooth underside and high ground clearance.

51

# Horses for courses
*Continued*

There's something in the Highland air—a nip, perhaps—that starts the Fling going. Mini-Moke and Haflinger perform a little dance.

## RENAULT 4L ESTATE

**Concessionaires:** Renault Ltd., Western Avenue, London, W.3.
**Specification:** 845 c.c. in-line four-cylinder o.h.v. water-cooled engine. Three-speed gearbox with synchromesh on all ratios. Front-wheel drive.
**Weight:** 11 cwt. 92 lb.
**Price:** £558 12s. 1d. (inc. p.t.).

THE advertising slogan "The remarkable Renault 4L" is well known and, believe me, it is justified. It is a *most* remarkable vehicle. The secret of its success seems to lie in the very sturdy box-member frame fitted with very soft torsion-bar suspension and a light body: the total weight is little over 11¾ cwt.

The "remarkable" feature is the manner in which it contemptuously gobbles up bumps; in a field with humps and gullies up to 12 inches in height and depth, it could be driven at 25 m.p.h. without any trace of juddering and no knocking sounds from the suspension. When it seemed that it *must* bottom its suspension on the far side of a jump it just didn't, but instead landed like a nicely brought down airliner.

For such conditions the unorthodox push-pull gearchange was ideal with no chance of missing the large knob or the desired ratio. Good low-speed engine torque was helpful at all times and, on hill starts, wheel slip on the greasy surfaces was the limiting factor. The clutch seemed indestructible.

Riding passenger in the 4L was highly popular during our tests because of its outstanding comfort and the astonishing amount of hot air which its heater blasted through.

As already mentioned, a slope which defeated it could often be conquered by going backwards, and reverse was high enough to permit all the speed which could be used in these conditions.

On the tarred road, the very soft suspension combined with the cross-country tyres to give rather a lot of wallowing and while cornering under power called for more and more turning of the steering wheel, lifting the foot caused the nose to tuck in neatly with a great feeling of security. This applied to icy corners as well.

The unsealed road was the place where the 4L really came into its own—not surprising considering that this is the sort of thing it is designed for—and it could be driven fast with practically no thought at all being given to the surface. And there was never any sound of engine or suspension working too hard.

Ideal for the small farmer, the 4L calls for the minimum of maintenance—the sealed radiator, suspension, steering joints and so on need no checking at all—and there is no doubt that its main road handling would be improved vastly with normal tyres. A full road test report appeared in our issue of December 30, 1964.

● ● ● ● ● ● ● ● ● ● ● ● ● ● ● ● ● ● ● ● ● ● ● ● ● ● ● ● ● ● ● ● ● ● ● ● ● ● ● ●

## MINI-MOKE

**Makers:** The Austin Motor Co. Ltd., Longbridge, Birmingham (also in Morris form).
**Specification:** 848 c.c. in-line, four-cylinder, water-cooled engine, mounted crosswise in unit with four-speed gearbox; synchromesh on three upper ratios. Front-wheel drive.
**Weight:** 10¼ cwt.
**Basic Price:** £407 (inc. p.t.). Extras include screen washer, sump guard, passenger seats, grab handles, etc.

FOR sheer fun, the Mini-Moke takes a lot of beating and it is not surprising that people are still thinking up new uses for it. It can be a beach car, estate transport, knock-about for surveyor or civil engineer, holiday camp taxi, baggage wagon, light personnel transport, general farm runabout and it is, above all, a source of delight to its driver.

On the rough ground it invited charging at hummocks to dive off the top of them and come crashing to earth on the other side. The suspension was such that the driver was never in any danger of being dislodged from his seat during this sort of antic. Hill climbing was limited by the length of level run available and wheelspin confined standing starts on grass to about one-in-four gradients.

I felt that really "hairy" tyre treads would have cured this as there was plenty of torque available, and the engine weight was in the right place. In snow and mud, quick side-to-side waggling of the front wheels often cleared a way and restored the missing traction.

Ground clearance could be another limiting factor in extreme circumstances yet at no time did I stick for this reason. Under more civilized conditions and treatment, the Moke invited full use of the accelerator in every gear, largely because of the predictable suspension behaviour and positive steering, which always did what it was told. The brakes, too, gave much peace of mind and on the unsealed road flat-out driving would have been quite a practical proposition.

On the tarmacadam the Moke was equally sure-footed and wheelspin deliberately provoked on the ice found it still following its nose in approved Mini fashion. Mud with a hard footing below gave it no trouble.

Although the comfort is limited largely to suspension, seats, windscreen and roof, there was very little draught inside although, of course, a side wind would alter this situation.

If I worked on a plantation, farm, nursery or any similar such place, I would certainly think up some good reason for owning a Moke: it is both odd and unfortunate that the authorities have seen fit to levy purchase tax on what is certainly a commercial vehicle in the widest sense. Even so, it comes out as good value and with side screens and heater fitted, its scope would be greatly increased.

Fuel consumption would depend entirely on the type of usage but 30 m.p.g. should be easy and 35 probable. **M**

# leisure

MOTOR CARAVAN TEST
(Extraordinary)

## 'Rover'

Richard Bensted-Smith tries French pioneering without tears in a **Land-Rover Dormobile four-berth motor-caravan**
(109 in. long-wheel-base: price £1273)

*Land-Rover Dormobile on site at Le Mans*

"HAWN", said Madame at the Camping of Le Portuais (in Anglo-Saxon you can get no nearer to the nasal honk by which the French express admiration), "C'est bon pour le Sahara." The Land-Rover Dormobile, I have no doubt, would indeed laugh at the desert and take most jungles in its stride; half the adventure, if not the fun, went out of exploring when amateur Livingstones were provided with such things. For a caravanning holiday in Brittany it has some imperfections compared with the more usual kind of Dormobile conversion which is commonly equipped with road springs, luggage space and provision for two people to stand up simultaneously, instead of taking turns. Beggars, however, do not get the option if they are taking a week off in the vehicle with which *Motor* has chosen to combine a road test and overnight shelter for its reporters at the Le Mans 24-hours.

There are in any case some compensating advantages. Even without the boat trailer (or, if you happened to have a large family, the trailer caravan) which a Land-Rover would whisk along behind it without a murmur, power is a handy thing to have on the Conti-

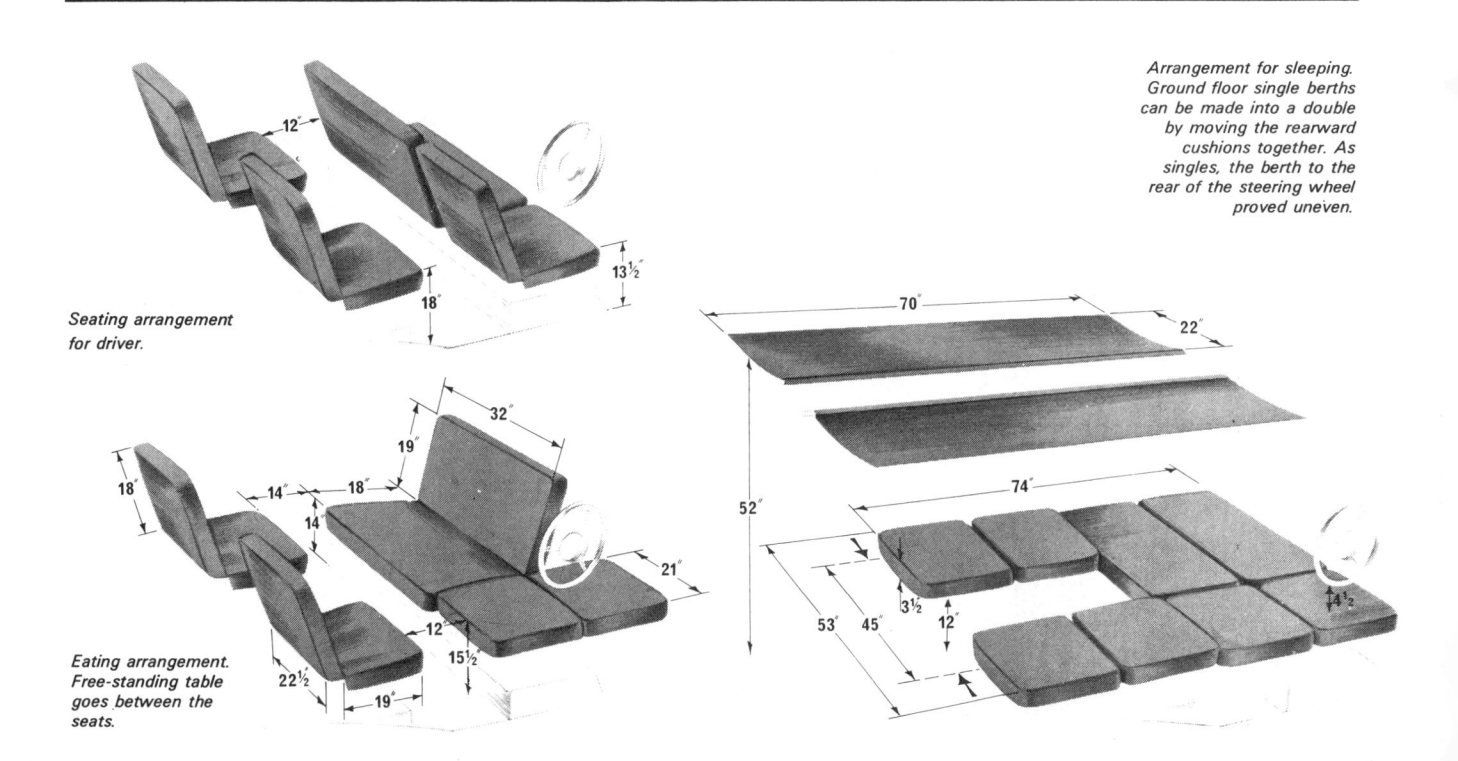

*Seating arrangement for driver.*

*Eating arrangement. Free-standing table goes between the seats.*

*Arrangement for sleeping. Ground floor single berths can be made into a double by moving the rearward cushions together. As singles, the berth to the rear of the steering wheel proved uneven.*

53

nent. So are a decent driving position, high ground clearance and four-wheel drive if the camp sites are muddy. Even more appealing at times is the moral influence of nearly two tons of Rover under full sail, seven feet tall and yawing slightly at 65 m.p.h. on a choppy Route Nationale: 2CVs preparing to hobble on to the main road in front of you slink back with a defeated look and you tell yourself that you're not really behaving like a tipper driver because the priority was legally yours anyway. It requires some strength of mind not to place your *pièce de résistance* in the path of oncoming traffic when you are emerging from a minor road.

We learned a certain amount about Rover the hard way, dodging the race traffic through country lanes around Le Mans. Driving from the top deck gives you a better view of imminent emergencies, but close study of the map is really possible only during the longer airborne moments or after you reach one of those supremely smooth-surfaced stretches of main road with which France is increasingly but still unpredictably dotted. On one such, as the domestic clatter from Rover's interior subsided, we turned up Section 8 of *Michelin Camping Caravanning en France* to see what the area of St. Malo had to offer.

Le camping (excuse me if you know it all already) is immeasurably better organized in France than here or in almost any other country. Michelin's symbols for camp sites, corresponding to the familiar peaked roofs of the hotel guide, are as you might expect a series of wigwams indicating anything from "rudimentary" to "extremely comfortable", with a description of the ground (grassy, slightly inclined), the rentals and a whole set of jolly little pictures to show how far you can take the pain out of living rough. Most of the two-wigwam establishments, only one removed from the bottom of the Michelin scale, offer at least cold showers, electric shaver points and a place for ironing laundry. At this level you're likely to pay about 1s. 4d. a head per night, plus 9d. for your car and the same amount for a tent or caravan. You are also likely, it is true, to encounter the sort of clean but physically taxing sanitation

*Bonnet mounting for spare wheel causes little obstruction but makes the driver uneasy. There is an alternative mounting on the tail-gate.*

*Les Pins camp site at Erquy at the western end of Brittany's Côte d'Emeraude. Rover has his top up.*

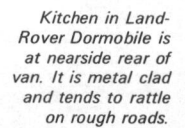

*Kitchen in Land-Rover Dormobile is at nearside rear of van. It is metal clad and tends to rattle on rough roads.*

which France has so far happily failed to export across the Channel, so it is usually worth forking out another shilling or so to move into a higher class.

Encouraged by an almost empty petrol tank and an urgent desire to visit a certain restaurant in Cancale, at the St. Malo end of the bay of Mont St. Michel, we renounced the pair of wigwams which was all Michelin could suggest in that vicinity. Instead we followed signs to the caravan-only site of Le Verger, poised dramatically on a clifftop within bracing distance of the wild waves and a fresh north-easter, expensive at 9s. a head but equipped to the point of luxury with everything from hot showers to a hose for washing down the car. We were, we told ourselves, getting run-in for greater rigours, and to make sure the shock would not be too severe we went off to the

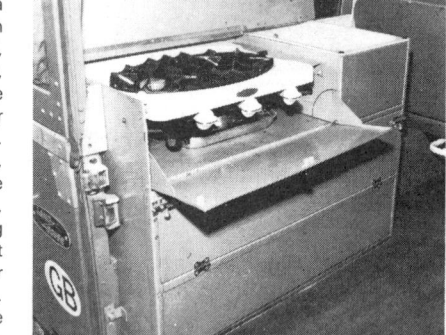

---

*View from the open tail-gate of the seats arranged to provide two single berths.*

### Performance

**Comfortable cruising speed:** 60 m.p.h.
**Top speed (mean of opposite runs):** 68.5 m.p.h.
**Acceleration, 0-50 through gears:** 20.7 sec.
**Standing ¼ mile:** 20.5 sec.
**Hill starting:** 1 in 4 gradient: Handbrake failed to hold but restart accomplished.
**Hill starting:** 1 in 3 gradient: Handbrake failed to hold but restart accomplished.
(4-wheel drive)
**Fuel consumption on tour:** 16.2 m.p.g. (2-star petrol)
**Speedometer:** Accurate at 50 m.p.h.; mileage recorder accurate.

### Specification

**Length:** 14 ft. 7 in. **Width:** 5 ft. 4 in.
**Height:** 7 ft. 1 in. (roof down).
**Ground clearance:** 9 in.
**Turning circle:** 45 ft. **Fuel capacity:** 16 gal.
**Engine:** 2286 c.c. 4 cyl., 67 b.h.p. net.
(**Note:** Future models will have 2625 c.c. 6 cyl. engine).
**Transmission:** Four speeds in two ranges giving eight forward ratios; two- or four-wheel drive. Synchromesh on third and top.
**Tyres:** 7.50 x 16, 6-ply.
**Extras:** Available for specialized use.

## leisure continued

*Foot marks in the sand show it to be soft, but 7.50 tyres and 4-wheel drive make Rover sure footed.*

*The front seat has to be removed to turn off the main gas cock which is rather inconvenient if you like to do this before shutting up for the night.*

Surcouf in Cancale to see if they still grilled lobster with a whisky sauce. They did, and the waiter was nice about serving a *"petit amusement"* of shrimps and winkles unasked-for with the aperitifs, and we were reminded as always that of the Loire wines Sancerre is not half as good as Brittany's own Muscadet to drink with fish. When the lobster didn't match up with the memory and the bill wildly exceeded it we decided that even if someone else had made off with the office copy of Michelin it still served us right for not spending our own money to be properly informed. Which is a sound principle but fallacious: the Surcouf still appears in Michelin 1967.

Back at Le Verger in windy darkness we listened thoughtfully to the waves, selected the lowest of Rover's eight gears (the handbrake is a powerful demonstration of faith with little strength left for anything else), raised the Dormobile roof and said some uncharitable things about the colleague who had evidently used the interior light switch as a step for reaching his top bunk the night before. Once you are up, these two "stretcher" berths are comfortably habitable, though rather less so on the starboard side where a heavy body makes the bunk sag enough to touch the side of the roof moulding. Below, too, the fold-flat seats make a better shaped bed on one side than on the other and we generally arranged ourselves one-up, one-down with the spare place on both levels used for baggage—a necessity in a crowded camp-site where you don't know the neighbours well enough to leave your best belongings outside.

You can live it up in the evenings very easily on the £50 allowance if you camp for the night, make your own French breakfast of a yard of bread and coffee, and have an Anglo-French picnic lunch of more yards of bread, paté, cheese and fruit. Really letting yourself go this way, with enough *vin ordinaire* to sink the thirst brought on by an arduous morning's round of the baker, grocer and pork-butcher, who may well be located several yards apart in the larger French villages, you will be hard put to it to spend more than eight shillings a head on lunch. At the Camping of Les Pins at Erquy, where we arrived the following night, breakfast was in fact a rather ravening affair after I had put my simple faith in the proprietor's assurance that the baker lived only 500 metres away, and had survived to pace it out as near three-quarters of a mile. The snag with French bread is that you cannot eat it more than 24 hours old (which makes Mondays a formidable problem) and the snag with a motor caravan is that you cannot very easily motor down to the shops before breakfast.

Erquy is near the western end of the *Côte d'Emeraude,* the rocky indented coast centred on the old Corsair city of St. Malo and the fashionable resort of Dinard which faces it across the Rance. Whatever the wind—even in a cold northerly blowing across the Channel—there is a fair chance of finding shelter somewhere along the tortuous outline of the Emerald Coast. Not surprisingly the whole northern coast of Brittany has a great deal in common with Cornwall—granite-based, fertile where its topsoil has not been swept away by Atlantic weather and cleft by deep, tidal inlets. The tides, if I'm not mistaken, have the greatest rise and fall in Europe—approaching 45 ft. at the highest springs—so that a tiny strip of beach can give way in six hours to a flat expanse of sand with the sea literally miles away. The story of flood tide in the bay of St. Michel coming in faster than a galloping horse is legend, but only just.

Having made no preliminary trials of Rover in 4-wheel drive against a galloping horse we didn't put the legend to the test but if you are carrying your house with you it is a comfort to be able to venture on to the sand when you get there. From Erquy you can attack the sea in a multitude of directions ("Temperated and iodizet climate", says the brochure, "recommended by Doctors. Picturesque and grandiose sites. Along nine miles of sea coast, SEVEN sandy beaches, differently turned"). Plugging gamely on down a jungle track past the camping of le Portuais, where we spent our third night for the sake of variety, we came out on a wild open spot remote from habitation and occupied only by a Simca Mille and an old Maigret-type Citroen, which took the keen edge off the pioneering spirit, but we also did a bit of exploring around some stranded boats on soft sand which would have had us sending up distress rockets from an ordinary motor caravan when the tide came in.

Both les Pins and le Portuais are three-wigwammers, adequately equipped for the washing of both bodies and clothes. I can recommend unravelling Gordian knots in a nylon clothes line as morning therapy for anyone who has dined well the night before. You can do that too, and cheaply, at the restaurant *à l'Abri des Flots* overlooking the fishing harbour in Erquy—so much so that after doing ourselves overproud on fishy things and Muscadet at 30 bob a head one night we went back the next night for more, and discovered only after a second memorable dinner that the chef had walked out in the interim. How many restaurant proprietors could step into the breach without their customers knowing?

You can have too much of a good thing and most especially the gastronomic reminiscenses of others, so I shan't labour our favourite subject except to urge that you try the Breton speciality of *crêpes,* or pancakes accompanied, as Michelin puts it by "cider or, for lovers of dairy produce, churned milk". I am not honestly mad about dairies but the *cidre bouché* which is dark, farmhouse cider with built-in champagne pop makes a splendid elevenses even without the pancakes.

You can, too (all right, I won't do it again), buy a shrimping net and devote some well-spent moments on the beach to catching little amusements as a preamble to the evening, bearing a couple of pints of English Channel back to boil them in and prove that you didn't bring a caravan with a cooker for nothing. We had them with a sundowner at le Portuais, and again in the super four-wigwam site of Cité d'Aleth at St. Servan overlooking the great tidal barrier power station which is being built across the Rance above St. Malo (you can drive across it now as an alternative to the ferry). And we had genuine Breton shrimps (in Breton seawater) on a Norman headland looking westward across a mile of sand to the last sunset of a fine, short week in June. We hit a deserted two-wigwammer that night with primitive plumbing and no electricity, the gastronomy caught up with my constitution and the next day the weather broke. but who cares? **M**

# TOUGH FOR THE ROUGH

*Both differentials turn at the same speeds ensuring positive front and rear drive, but the wheels either side share the torque and grip or slip according to load and surface*

## LAND-ROVER ON ROAD AND TRACK

FROM the factory at Solihull, Land-Rovers have now been going out to all parts of the world for 16 years; and the manufacturers must long ago have learnt practically all there is to know about cross-country vehicles. Certainly the Land-Rover has won a remarkable reputation for reliability and strength, and it has become the automatic choice for trail-blazers heading for remote areas of the globe.

In its square panelled and ultra-functional-looking exterior there are not many changes to be spotted between today's product and the original version, for the styling is ageless. No doubt the Land-Rover of five or 10 years ahead will still look exactly the same. Mechanical changes are also rare, and the specification has remained unaltered since the 4-cylinder overhead-valve engine, originally used in the Rover 80, replaced the previous L-head engine. Its compression ratio is only 7 to 1 so that there is never need to buy anything but the cheapest petrol available, and it develops 77 (gross) b.h.p. at 4,250 r.p.m. Its swept volume is 2,286 c.c., and there is a diesel alternative of the same size, developing 62 b.h.p.

Taken up to high revs in indirect gears, there is a throaty, businesslike roar from the engine—but none of the sort of mechanical clatter which would discourage one from working the engine hard. Often when overtaking it pays to hold

## Performance

### ACCELERATION
| | |
|---|---|
| 0 to 30 m.p.h. | 6.9 sec |
| 0 to 40 m.p.h. | 12.3 sec |
| 0 to 50 m.p.h. | 19.9 sec |
| 0 to 60 m.p.h. | 36.1 sec |
| Standing quarter-mile | 24.0 sec |
| 10 to 30 m.p.h. (top gear) | 10.2 sec |
| 20 to 40 m.p.h. (top gear) | 13.8 sec |
| 30 to 50 m.p.h. (top gear) | 13.2 sec |
| 40 to 60 m.p.h. (top gear) | 22.8 sec |

### MAXIMUM SPEEDS IN GEARS
| | M.p.h. | |
|---|---|---|
| Top | 67 | (mean) |
| | 69 | (best) |
| Third | 56 | |
| Second | 41 | |
| First | 26 | |

### MAXIMUM SPEEDS IN LOW RATIO
| | M.p.h. |
|---|---|
| Top | 33 |
| Third | 22 |
| Second | 15 |
| First | 10 |

### BRAKES (at 30 m.p.h. in neutral)
| Pedal load in lb | Retardation | Equiv. stopping distance |
|---|---|---|
| 25 | 0.25g | 120 |
| 50 | 0.50g | 60 |
| 75 | 0.72g | 42 |
| 100 | 0.82g | 37 |
| 125 | 0.90g | 33.4 |
| Handbrake | 0.34g | 89 |

**BATTERY:** 12-volt, 57-amp. hr.

### WEIGHT
| | |
|---|---|
| With half-full petrol tank | 26.9 cwt (3,010 lb) |
| Distribution (per cent) | F, 55.3; R, 44.7 |

### DIMENSIONS
| | |
|---|---|
| Height | 6ft 5½in. |
| Headroom, interior | 4ft 0½in. |
| Width | 5ft 6in. |
| Length | 11ft 10½in |

### PETROL CONSUMPTION
Grade: Regular
Overall, for 878 miles including cross-country running .. 18.3 m.p.g
Normal range .. 15-22 m.p.g

---

third gear to at least 50 m.p.h. At the other end of the scale the power unit is also very smooth, and will pull away without snatch from 10 m.p.h. in top gear. The 67 m.p.h. maximum speed is fixed by power limitations and the high wind resistance of the vehicle's uncompromising shape. Descending long gradients, the gearing allows slightly faster speeds still without over-revving.

A very quiet starter motor, which gives the engine a really vigorous impulse, seldom needs to be used more than once, nor is there much need for the choke in mild weather. The engine's very steady, even tickover is a good feature for prolonged static running when the optional power take-offs are in use.

Although it has a high seating position, giving a commanding view over other traffic, the blind sides of the canopy are very restricting to the rear quarters, and one is never quite sure where the front corners are. Bearing in mind the serious damage which its huge girder bumper could do in the confines of a car park, one has to proceed with special care when manœuvring, though there is a safe tendency to underestimate and think that the clearance is much less than it really is.

Quite adequate acceleration, with a time from rest to 30 m.p.h. of 6.9 sec, makes it easy to keep the Land-Rover well up with the traffic stream. A clutch with real bite and a positive, sturdy gear change, all help the driver make brisk

→

*Left: The engine is well protected; it stays clean after repeated high speed water splash duckings and there was never any trouble with damp on the "electrics" Right: Functional but unbeautiful is the key to the interior. The optional hand throttle was fitted to the test car. Seat upholstery is in washable p.v.c., also used for the door trim*

*From the beginning, the Land-Rover has been faithful to beam axles and half-elliptic leaf springs for its sturdy suspension. It has 10in. ground clearance unladen, and very long vertical wheel travel is allowed by the suspension*

## TOUGH FOR THE ROUGH...

progress, but synchromesh on second gear was weak on one of the two vehicles tried.

Earlier, a member of the staff took a short-wheelbase model with enclosed cab to Wales for an extended trip. Tried on the road in this way, the Land-Rover is reasonably quick and manageable, but its drawbacks are a very joggy and firm suspension—much like the ride associated with the older kind of sports car—and rather vague steering. Directional stability is not good, and the car always seems to be trying to wander off at a tangent. The steering is also heavy at low speeds.

A common complaint about the Land-Rover is of the very upright seating posture, and the near-vertical action of the pedals. The sideview silhouette of a Land-Rover driver has the same right-angle shape as that of a starting handle and rather acute ankle angles result. On a long trip this can be tiring, but, as familiarity with the vehicle increases, it is noticed less, and it does not restrict the effort which needs to be exerted on the brake pedal for good response.

With 75 lb load, the brakes pull strongly enough to lock the rear wheels when there is no load on board, but an even harder 100 lb shove on the pedal results in more efficiency. Still heavier braking—125 lb on the pedal—gave the Land-Rover's best possible result of 0·9g, which is good for this sort of vehicle running on compromise cross-country tyres. The handbrake works on the output shaft from the transfer gearbox, and is really effective. A firm pull holds it securely on some of the incredibly steep hills which can be climbed—almost wherever the wheels can get sufficient grip.

After using the Land-Rover much as a car or light van for some days, the limitations mentioned, and especially the fuel consumption for what is essentially a three-seater, showed that it is rather out-of-place as an ordinary traffic vehicle. However, after covering a really difficult test course of steep tracks and rough ground, we soon formed very different opinions.

Only after prolonged practice, and perhaps one or two spills, would one learn just how far to go with a Land-Rover and not overdo it. Often we would survey a test hill from the bottom, shake our heads, and then finally pluck up courage to have a crack at it. Then, with the engine roaring as we dropped down through the gears, the stage would be reached when the Land-Rover had practically come to rest with full throttle in bottom gear, and seemed to be standing practically vertically; there would be an awful moment of wondering what might happen if it did not make it. Just before momentum is lost, down goes the clutch and the red-knobbed lever of the transfer gearbox is jerked back with a slight crunch of gears, and suddenly the vehicle is away again, climbing at a terrifying angle. We came to the conclusion that its hill-climbing abilities exceeded our confidence and courage, and that it would surmount anything which, even in a daring mood, we were prepared to venture.

Similarly, the suspension and steering, which have seemed too strong and heavy for main roads, are felt in quite a different vein when pounding over the rough. Huge dips and gullies are swallowed up extremely well by the large range of vertical movement, and although the ride is decidedly harsh, so that the passenger has to cling to his seat and the driver to hold hard to the steering wheel, magnitude of the bumps which can be tackled at speed is remarkable. Again, the abilities of the vehicle tend to exceed those of its crew, and the limitations of how much pounding and bouncing one can take are reached long before hitting anything hard enough to damage the structure. No safety belts were fitted; we do advocate them for keeping cross-country vehicle crews in their seats, but the seats themselves are comfortably "dead" and do not accentuate the bounce, nor are they hard to land on when one returns to base after a particularly bad bump.

Some reaction comes back through the steering, and obviously fairly vigorous work at the wheel is needed to keep on course when hammering over rough ground. Control is easy, and there is no violent kick-back, nor is the steering at all heavy or over-sensitive in these conditions.

During our tests with this Land-Rover the ground was dry, and whatever mud we could find was little test for it. From past experience with these vehicles we know that most conditions likely to be encountered, whether snow or mud, can be tackled provided they are attacked with sufficient verve. One must build up speed at all costs and get into the right gear, and then keep going hard. A moment's pause because it looks worse still ahead, or to

*A big oil bath air cleaner is standard. All filler caps and the dipstick are accessible; the front bumper forms a good step for anyone making engine adjustments*

Short-wheelbase (88in.) Land-Rover Regular manufactured by The Rover Co. Ltd., Solihull, Warwickshire.

**PRICES:**

|  | £ | s | d |
|---|---|---|---|
| Land-Rover Regular as tested | 691 | 0 | 0 |
| Regular with diesel engine | 805 | 0 | 0 |
| 109in. l.w.b. model | 790 | 0 | 0 |
| 109in. with diesel engine | 904 | 0 | 0 |

**EXTRAS:**

|  | £ | s | d |
|---|---|---|---|
| Detachable hardtop | 40 | 2 | 9 |
| Front capstan winch | 46 | 0 | 0 |
| Passenger windscreen wiper | 4 | 3 | 0 |
| Heater and demister | 12 | 0 | 0 |
| Free-wheeling front hubs | 22 | 10 | 0 |
| Interior rear mirror |  | 7 | 8 |
| Power take-off, rear | 41 | 17 | 6 |
| Power take-off, centre | 14 | 0 | 0 |
| Rear seats for two (pair) | 13 | 5 | 0 |
| Water thermometer and oil pressure gauge | 6 | 16 | 0 |
| Windscreen washer | 1 | 17 | 6 |

change gear, can spoil the run. If these techniques are used, it has to be deep clogging mud or a very steep sand dune before a Land-Rover will get stuck and need assistance.

For such conditions it is a great help that four-wheel drive can be engaged now on the move, simply by pressing on the yellow-knobbed lever protruding from the top of the transfer gearbox. On earlier models it used to be necessary to stop, engage low ratio and then, while holding the four-wheel drive knob down, push the level back to direct drive. Four-wheel drive would then stay engaged, but this made it much more complicated to engage four-wheel drive in the direct ratio, and the tendency was to try and tackle most conditions in two-wheel drive. Now the four-wheel drive lever can be banged down to add the extra traction of the front differential whenever doubtful soft ground is spotted. Four-wheel drive should not be used on the hard, of course, because of transmission wind-up and a resultant increase in tyre wear.

It seems a pity that the utility nature of the vehicle should be emphasized by rather crude facia layout and interior fittings, with ugly windscreen wiper motors, heater pipes and angular presswork all revealed. Even new lorry cabs are beginning to look less basic than this nowadays. The heater is an extra, and is only a recirculatory unit just sufficient to keep the front compartment reasonably warm, and the windscreens demisted, in winter. The lamps are well recessed away from frontal damage when ploughing through undergrowth, and there are winking indicators with a column-mounted switch incorporating its own green warning lamp. Cool air vents are standard at the base of each windscreen.

For traversing ground where there are no proper roads a whole variety of vehicles are available, but only one—the Austin Gipsy—is, by its similar construction and character, comparable with the Land-Rover; and inevitably comparisons are always made between the two. In our recent test of a Gipsy (25 June) a glass fibre hardtop was fitted to the test vehicle, but this could not entirely account for its weight excess of 3cwt over the Land-Rover. With less weight and more power from its slightly bigger engine, the Land-Rover is understandably quicker on acceleration, but it is a creditable reflection of its efficiency that it is more economical as well.

Regularly during the test it exceeded 18 m.p.g.—good for this kind of cross-country vehicle, as a fair amount of slogging off the beaten track was included. A much larger fuel tank is needed than the 10-gallon one fitted; long wheelbase models have a 16-gallon tank, offering a more practicable range. To assist refuelling in the field, the filler has a large diameter orifice with a pull-out extension incorporating its own filter. The filler cap is recessed into the flat side of the vehicle to protect it from damage.

There is certainly room for improvements in the Land-Rover's comfort for use on ordinary roads, while its utility appearance might be modified without necessarily detracting from its very functional character; but these are small points to set against the vehicle's remarkable cross-country performance. ■

*Drawbars for casual towing are fixed either side at the rear. It is quite an elaborate performance to release the strings and webbing to open the rear canopy, enabling the tailgate to be dropped down*

# MORE POWERFUL LAND-ROVER

SIX-CYLINDER ENGINE AVAILABLE FOR LONG WHEELBASE MODEL; MINOR INTERIOR IMPROVEMENTS

LONG wheel-base versions of the Land-Rover are now available with the six-cylinder 2·6-litre power unit which is used in the forward control model. The engine is based on that once used in the Rover 100 saloon, and has compression of 7·8 to 1. Maximum power is 85 b.h.p. net and peak torque is 132lb. ft. at 1,500 r.p.m. For markets with low octane fuels, a 7-to-1 compression is offered with max. power of 81 b.h.p. and peak torque of 128lb. ft.

With the six-cylinder versions, wider front brakes are fitted and a servo is standard. Lining area is increased by 12 per cent. The battery is repositioned under the left seat, and the tool box is moved to a new position under the centre seat at the front. The dynamo has higher rate of charge, and the same carburettor air cleaner of improved design as is fitted to the forward control Land-Rover is used.

The bigger engine gives much more performance and low-speed torque than the four-cylinder, and performance figures taken fully laden, by the makers, claim a reduction of more than 10 sec in the acceleration time from rest to 50 m.p.h., which takes just over 20 sec. A 90 m.p.h. speedometer is fitted.

As well as these improvements, which apply only to the new six-cylinder model, some general modifications have been made on the whole range. Instruments have been restyled, and the minor controls are rearranged for neater appearance on the central facia panel. The old separate starter switch is replaced by a combined key-operated ignition and starter switch. The handbrake has been extended and now comes closer to the gear lever, so that it can be reached more easily by a driver wearing seat belts. The old and rather clumsy arrangement of separate wiper motors with individual controls is replaced by a single wiper motor driving twin blades through a concealed rack at the base of the screen, which can still be folded flat on some models as before. The electrical system now employs a negative earth.

Well over half a million Land-Rovers have now been produced, of which more than 70 per cent have been exported to a total of 65 countries throughout the world. This four-wheel-drive all-purpose vehicle was originally introduced in 1948 and was powered by a four-cylinder engine of 1,595 c.c., developing 50 b.h.p. Capacity was increased to 2-litres in 1952, and the wheelbase was extended by 6in. in 1954, with the addition of a longer wheelbase optional model two years later. The change to the current 88in. and 109in. models with

| PRICE LIST | | | BRIEF SPECIFICATION | |
|---|---|---|---|---|
| | 4-Cyl. £ | 6-Cyl. £ | Cylinders | 6, in-line |
| 88in. Short Land-Rover .. | 710 | — | Cooling system | Water; pump, fan and thermostat |
| 88in. Station Wagon .. | 860 | — | Bore | 77·8 mm (3·063in.) |
| 109in. Long Land-Rover .. | 813 | 873 | Stroke | 92 mm (3·625 in.) |
| 109in. Long Land-Rover Deluxe .. .. .. | 833 | 893 | Displacement | 2,625 c.c. (161 cu. in.) |
| | | | Valve gear | Overhead inlet, side exhaust |
| 109in. 10-seater Station Wagon .. .. .. | 1,235 (inc. tax) | 1,308 (inc. tax) | Compression ratio | 7·8 to 1: Option, 7 to 1 |
| | | | Carburettor | One SU HD6 |
| | | | Fuel pump | SU electric |
| 109in. 12-Seater Station Wagon .. .. .. | 1,013 | 1,073 | Max. power | 85 b.h.p. (net) at 4,500 r.p.m. |
| | | | Max. torque | 132 lb. ft. (net) at 1,500 r.p.m. |
| | | | Max. b.m.e.p. | 124 p.s.i. at 1,500 r.p.m. |

*Above: The new Land-Rover interior has also been applied to the established 4-cylinder versions. Note the re-styled instruments and the key starter. Below: The lengthy 6-cylinder engine fits neatly into the bonnet space with room to spare. The 2·6-litre power unit develops 85 b.h.p. (net) in this application*

2¼-litre four-cylinder engine was made in 1958 and the 250,000th Land-Rover was turned out in the following year (after 11 years of production). The forward control model, also powered by the six-cylinder engine, was added to the range in 1962.

With the option of the 6-cylinder engine, Land-Rover buyers can now choose from three engines (including the diesel), two wheelbase lengths, and no fewer than 38 body styles.

Prices of four-cylinder models are unchanged. The extra cost of the six-cylinder engine and associated improvements is £60 (£73 including tax on the 10-seater Estate Car).

# LAND-ROVER SIX

Road and Cross-country Tests of the 12-Seater 2.6-Litre Estate Car

WITH the introduction in April of the 6-cylinder 2.6-litre power unit as an option for the long wheelbase Land-Rover models, the Rover Company offered an even wider choice, and there is now an excellent range of Land-Rovers available. From the basic short wheelbase with canvas canopy it goes right through to the big, forward-control model; which is more or less a Land-Rover lorry. The 6-cylinder engine was previously available only for the forward-control version, on which it is the standard power unit. In the ordinary long wheelbase models it offers useful extra power and torque for both hard cross-country work and higher speeds on the road. The 6-cylinder unit is not available for short wheelbase models. It is a long-stroke engine of 2,625 c.c. capacity.

In this country the purchase tax regulations affecting these vehicles are quite illogical. The versions with canvas tilt to the rear of the cab escape tax because of their "commercial" nature and the lack of side windows to the rear of the driver's seat. The estate car, however, is regarded as a private vehicle and is subject to purchase tax unless 12 seats are fitted. It is then regarded more as a small bus, and again purchase tax is waived. This has the rather ludicrous result that instead of costing more, the extra two seats actually give a net saving of £235. All models, however, escape the 40 m.p.h. commercial speed limit because 4-wheel drive classifies them as dual-purpose vehicles. Somewhere in all this, we feel sure, there must be some logic but the authorities have yet to point it out to us. The outcome is that the tax-free 12-seater is a particularly popular model in this country when the estate car is wanted instead of one of the canopy models, and one of these forms the subject of this test.

**More Power for Hills**

Those familiar with the Land-Rover will find the extra torque of the bigger engine most noticeable in the ability of this heavy vehicle to climb long hills pulling strongly in top gear, where one of the 2¼-litre four-cylinder Land-Rovers would almost certainly have needed to drop down to third. The engine is a very prompt starter and invariably responds to the first turn of the key; and it is scarcely audible from inside when idling. A new refinement on all models is that the separate starter button beneath the facia is replaced by a combined ignition and starter switch, worked by the key in the way which is now (Jaguar and Daimler excepted) almost universal. Little choke is needed and the mixture control can be pushed home long before the thermostatically-controlled warning light comes on.

Once the car is on the move, the noise level of the engine increases a lot, and from being so quiet at tickover it produces considerable power roar which makes one feel that about 50 m.p.h. is fast enough. In fact, the noise level does not then get much worse if the speed is increased right up to the maximum of just over 70 m.p.h. The noise heard is a purposeful sound without any mechanical clatter or signs of distress, and the driver soon becomes accustomed to it and turns a deaf ear.

The estate car is more than half a ton heavier than the short wheelbase canvas-top model which we tested in November 1965—the 6-cylinder unit has an aluminium cylinder head and weighs hardly any more than the 4-cylinder. In spite of this 10cwt penalty the 6-cylinder estate car is a lot quicker, and accelerates from rest to 60 m.p.h. in 29 sec compared with a time of 36·1 sec for the four-cylinder s.w.b. version. The same gear ratios are used with either engine and regardless of wheelbase, but although the 6-cylinder unit develops its maximum power of 90 (gross) b.h.p. at 4,500 r.p.m., against 77 b.h.p. at 4,250 r.p.m. for the 4-cylinder, maximum speeds in the gears are a little lower. In third gear, we obtained 56 m.p.h. with the 4-cylinder Land-Rover, but the 6-cylinder sounded as though it had reached its limit at 53 m.p.h.

Indirect gears are well spaced, and the light but very positive gear change, with rather long travel, seems particularly appropriate to a heavy duty vehicle of this kind. It becomes natural practice to change gear rather slowly and definitely with double-declutching both up and down although this is not entirely necessary. Even with no synchromesh on first or second gears, changes into these ratios are made without crashing the gears provided the movement is not hurried. In acceleration testing, of course, there was a nasty crunch of gears each time the lever was pulled smartly back into second. The clutch takes up smoothly over quite a long range of travel, but is rather heavy for repeated use in traffic.

Naturally this big vehicle is seen at worst in congested conditions when its considerable size and rather blind front corners ▶

# LAND-ROVER . . .

are an embarrassment. The steering at low speeds is heavy and the turning circle is so big that it is difficult to manoeuvre the vehicle. Although the driver sits high and has a commanding view with the ability to see over the top of most cars, he tends to forget that the flush sides save width, and often thinks that the business-like front bumper sticks out more at the front than it really does, so there is a considerable safety margin.

It is when wide open spaces are available, and on tracks which would be impassable to ordinary traffic, that the Land-Rover really shows its mettle and our usual test circuit over the Aldershot fighting vehicles terrain reaffirmed our earlier praise for the Land-Rover's abilities. The suspension is very hard, and on ordinary roads the ride is far too bumpy for comfort, but when pounding over really rough going it absorbs huge bumps and potholes impressively well. Occupants have to hold on hard to prevent being thrown about, but one never has any fear of hitting a ridge or ditch too fast and damaging the vehicle. Its ground clearance of nearly 10 in. enables it to climb over the sort of humps and gulleys encountered in real cross-country driving without fear of scraping the chassis.

Early in our colonial tests the left wheels dropped into a deep ditch concealed by the undergrowth. This is the surest way to get stuck even with a 4-wheel drive vehicle, because the weight is transferred from the wheels which are on the hard and have good grip, to the side which is bogged down. The angle of the vehicle seemed quite alarming from inside although well within the 45 deg. safety limit before it will roll over. By some patient to and fro work using 4-wheel drive and the transfer gearbox, the Land-Rover eventually and very impressively climbed out.

Pitting the vehicle against what seemed near-vertical hills showed dramatically how well it will climb in the lowest of its eight available ratios, and it came as quite a relief when the wheels eventually lost their grip on the rough gravel and it would not climb any farther.

Cross-country controls are the same with the 6-cylinder as with previous models. A spring-loaded vertical lever beside the gear lever is pushed downwards to bring drive to the front wheels in addition to rear wheel drive, and this can be done while on the move, when sticky conditions are seen ahead. If the driver is already preoccupied at the steering and needs both hands he can work this lever by stamping on it with his left foot. For very exacting work, such as climbing a really steep hill or tackling a ditch, the red lever to the right of the gearbox is pulled backwards through a neutral position to engage the low ratio of the transfer gearbox. This automatically brings in 4-wheel drive as well, so that the torque multiplication is shared by both differentials.

In first gear with the low ratio of the transfer gearbox engaged, the overall ratio is nearly 40 to 1, and in top gear in this condition the ratio is roughly the same as that of second gear when the transfer gearbox is not in use. As soon as reasonable going is available again, the lever can be moved back to the normal direct-drive position, and it is important to remember that to go into or out of the transfer gearbox low ratio, the vehicle must be brought to rest. To revert to ordinary 2-wheel drive one simply engages and then disengages the low ratio, so it is again necessary to stop in order to get out of 4-wheel drive, although, as explained, 4-wheel drive may be brought into action while on the move. Excessive tyre wear and wind-up would result if 4-wheel drive were left in use on hard surfaces.

In two respects the new 6-cylinder seemed less satisfactory than earlier models we have tried. The first fault was the ease with which the ignition was flooded after taking a deep water splash at even quite moderate speeds. Water seems to be thrown up against the bonnet and showers down over the distributor, flooding the electrics. Each time, it ran unevenly, with audible tracking, until the moisture dried. The other weak point was the ease with which exhaust fumes found their way into the interior. With the six-cylinder estate car the exhaust protrudes through a hole in the rear mud flap on the left side of the vehicle, and fumes seem to be sucked into the car past the sealing of the tail gate. By opening the vents below the windscreen and small extractors in the roof, a good flow of fresh air through the vehicle was obtained to help reduce it.

Included with the extra cost of £60 for the 6-cylinder engine is the addition of a servo for the drum brakes, and lining area at the front is increased by 12 per cent. The

*All the sliding side windows have locking catches, and door locks and fold-down steps are also standard with the 12-seater*

| ACCELERATION | | |
|---|---|---|
| 0 to 30 m.p.h. | | 5.8 sec |
| 0 to 40 m.p.h. | | 11.3 sec |
| 0 to 50 m.p.h. | | 17.1 sec |
| 0 to 60 m.p.h. | | 29.0 sec |
| Standing quarter-mile | | 23.6 sec |
| 2nd gear: | | |
| 10 to 30 m.p.h. | | 5.3 sec |
| 3rd gear: | | |
| 10 to 30 m.p.h. | | 7.8 sec |
| 20 to 40 m.p.h. | | 8.0 sec |
| 30 to 50 m.p.h. | | 9.8 sec |
| Top gear: | | |
| 10 to 30 m.p.h. | | 11.3 sec |
| 20 to 40 m.p.h. | | 13.1 sec |
| 30 to 50 m.p.h. | | 15.8 sec |
| 40 to 60 m.p.h. | | 19.5 sec |

| MAXIMUM SPEEDS IN GEARS | M.p.h. | |
|---|---|---|
| Top | 73 | (mean) |
|  | 75 | (best) |
| 3rd | 53 | |
| 2nd | 36 | |
| 1st | 22 | |

| MAXIMUM SPEEDS IN LOW RATIO | M.p.h. |
|---|---|
| Top | 39 |
| 3rd | 26 |
| 2nd | 18 |
| 1st | 11 |

**BRAKES** (at 30 m.p.h. in neutral)

| Pedal load in lb | Retardation | Equiv. stopping distance |
|---|---|---|
| 50 | 0.40g | 75 |
| 75 | 0.70g | 43 |
| 100 | 0.95g | 31.7 |

**WEIGHT**

With half-full petrol tank .. 35.2 cwt (3,948 lb)
Distribution (per cent) .. F, 45.3; R, 54.7

**DIMENSIONS**

| Height | 6ft 9in |
|---|---|
| Headroom, interior | 4ft |
| Width | 5ft 6in. |
| Length | 14ft 7in |

**PETROL CONSUMPTION**

Grade: Regular 2-star
Overall, for 576 miles including cross-country work  13.8 m.p.g.
Normal range .. .. .. .. 12-15 m.p.g.

---

brakes feel much the same as on the short wheelbase model, in spite of the greater weight of the long wheelbase estate car, and maximum efficiency is appreciably higher, with a commendable 0.95 g available in return for 100 lb load on the pedal. This is exceptionally good for a vehicle on coarse tread tyres, but not surprisingly two thick black lines were left on the road. The handbrake works on the rear transmission and cannot be used on the move without causing severe vibration. As a parking brake it is terrifically strong and will hold the Land-Rover securely on the steepest gradient it is able to climb. The brake is now located beside the driver's left knee and is much more easily reached than before.

In the improvements to the interior, the clumsy separate wiper motors have been replaced by paired blades driven from a concealed rack at the base of the windscreen, with single on-off switch between the instruments. They are self-parking and sweep good arcs, but we feel the next improvement on the Land-Rover should be the addition of foot-operated screen washers. When ploughing through muddy water at speed a lot of the splash is flung forward on to the windscreen, and washers would be invaluable. The new 90 m.p.h. speedometer is very clear to read and proved exactly accurate right up to 70 m.p.h. In the matching instrument on the left are temperature and fuel gauges and an ignition warning lamp. An oil pressure warning lamp is incorporated

*Just visible beyond the seat (left) is the push-down knob for adding drive to the front axle. The lever to its right is the transfer gearbox control, which also introduces four-wheel drive. Right: There is ample space for the six-cylinder engine and access for service is still quite good*

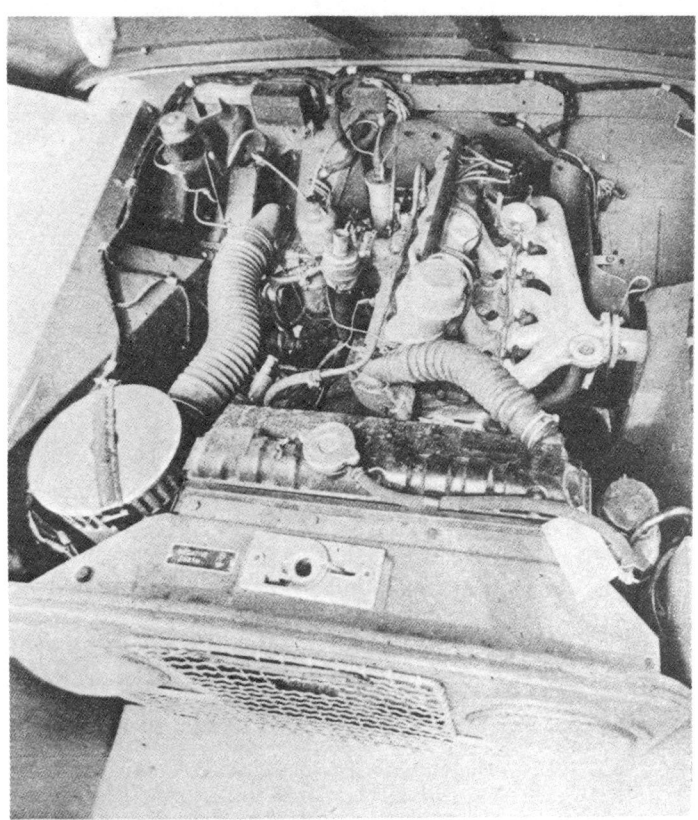

# LAND-ROVER . . .

in the bottom of the speedometer, with the rich mixture telltale and headlamps main beam indicator. The direction indicators are self-cancelling and the switch includes its own green warning lamp.

The few refinements have made the interior look less untidy than before, without detracting from its essentially utilitarian character. In the 12-seater estate car, the row of seats behind the driver can be folded forward to give access to the pair of longitudinal seats at the back, or people can climb in quite easily through the tail opening. Door struts would be appreciated to hold the doors open on a slight incline, and the driver also has to remember to check that the folding steps have been kicked back to the vertical after use.

Tried with its full complement of 12 people on board, the Land-Rover handled extremely well, and although it rolls a great deal on corners, which is particularly noticed by those who sit facing each other at the rear, it remains very stable. The extra weight of full load takes all the harshness out of the suspension without introducing any tendency to pitch.

Regular grade fuel is adequate for the 7·8 to 1 compression ratio, and a 7 to 1 ratio is available which should enable the engine to run satisfactorily in countries where only the poorest, low octane fuel is on sale. Up to 15 m.p.g. is possible if not too much use is made of the cross-country abilities, and the fuel tank holds 16 gallons. A pull-out extension to the filler enables one to refuel in the field from a can without spillage, and includes a filter.

The revised Land-Rover is still very much a utility vehicle, specifically designed for hard use and exacting conditions. The 6-cylinder engine option is well worth its extra cost for the much improved hill climbing and better performance which results.

*It is in really severe cross-country conditions that the Land-Rover excels, and it was able to extricate itself from the ditching (above, left); but improved water-proofing is needed for conditions as above, right. Below: a demonstration of ground clearance*

*Long wheelbase (109in.) Land-Rover 12-seater estate car, manufactured by The Rover Company Ltd., Solihull, Warwickshire.*

**PRICES** £ s d

| | |
|---|---|
| 109in. 12-seater estate car with 6-cylinder engine (no purchase tax) | 1,073 0 0 |
| Lap and diagonal safety belts (each) | 2 7 6 |
| Heater and demister | 12 11 0 |
| Rubber pedal pads | 8 0 |
| Sun visors (each) | 10 6 |
| Windscreen washer | 1 17 0 |

SINCE introduction in 1948, Land-Rovers have been progressively improved, although appearance has not altered much, and the range of specialised equipment available has been extended enormously. Here we take a look at the present range of equipment and the development of the vehicle.

IT was the post-war shortage of steel which started it all. Steel became rationed, like everything else, and it was allocated to our industries in proportion to the value of their exports. At the time, Rover's fairly expensive quality cars were not selling well abroad, and a new venture was needed to attract overseas buyers. The answer chosen was to build a jeep-type vehicle on a commercial basis, and the first Land-Rover prototypes were built during 1947. The project was launched in the following year at the Amsterdam Motor Show.

Now that 20 years of uninterrupted production have been completed, Rover claim that 600,000 Land-Rovers are operating in more than 170 different territories all over the world. In the early days you only had to make a noise like a round-the-world trip and you could borrow one, chalked up to publicity and development; but the product was soon proved and established all over the world, and the free loans ended. Demand has constantly outpaced supply, and whenever there is a recession in car sales, Rover are able to divert their attention to catching up on the backlog of Land-Rover orders. The utilitarian vehicle born in hard times has certainly earned for Britain some £270 million in foreign exchange. Only last week an order for 74 Land-Rovers in Brazil was announced, worth £70,000.

Outward appearances have changed remarkably little, but major developments, particularly of the power unit, have been introduced over the years; a concise history is given in the Used Car Test in this issue. Recent introduction of the six-cylinder engine extends the range available, and there are now five basic Land-Rovers (including the forward control model), and 38 body styles. On the following pages we show just a glimpse of the range of applications and special equipment.

On entering a Land-Rover the first impression is of the great height off the ground, which gives an eye level well above the roof of most cars. The angular front wings are in view, but allowance has to be made for the sturdy bumper out front, and of course it is not a question of what damage the vehicle will suffer so much as what it will do to anything it hits. They are made for an outdoor life, washed by the rain and occasionally hosed out to clean the interior. There are two little chrome rims round the headlamps—the only concession to styling; everything else, practically, is either galvanised steel (including the chassis) or painted aluminium.

The four cylinder petrol engine has a 2¼-litre capacity, three main bearings and 7-to-1 compression ratio; it develops 77 bhp (gross) at 4,250 rpm, and was derived from the Rover 80 car engine. The diesel unit has the same 2,286 c.c. capacity, but 23-to-1 compression and develops 67 bhp (gross) at 4,000 rpm. The 6-cylinder, available only on long wheelbase models, derived from the Rover 100 car engine, has a 2,625 c.c. capacity, long-stroke instead of being over-square (as the four-cylinder), and there are alternative compression ratios—7 or 7.8 to 1. Gross power outputs are 91 bhp for the lower compression unit, or 95 bhp for the higher, both of which accept 2-star regular grade fuel. This is a seven-main-bearing engine, with overhead inlet and side exhaust valves, and there is no diesel version of the six-cylinder.

Probably the reliability of the vehicle owes much to its very simple, well-tried layout. Both axles are live, and longitudinal leaf springs are used with telescopic dampers. The engine is mounted fairly high in the chassis and drives through a combined gearbox and transfer gearbox unit to the rear wheels or to all four wheels. Behind the gearbox is a drum brake ahead of the prop shaft to the rear transmission, for parking only; ordinary braking is by drum brakes front and rear, with servo assistance only on the forward control model.

In town Land-Rovers are a bit clumsy and unmanoeuvrable, and on the open road they are relatively slow. It is, of course, in cross-country ability, combined with acceptable performance on the road, that they really score. It is quite remarkable how well a Land-Rover will keep going in only two-wheel drive, and the suspension which seems so needlessly harsh on the road, proves its worth in the way in which it absorbs monstrous gulleys and potholes without any need to reduce speed. The huge ground clearance pays off, too, 8in. on the 88in. wheelbase Regular and 9¾in. on the 109in. wheelbase Long model.

When at last it does get stuck, or when the driver sees really soft ground ahead, he has only to kick down on the yellow knob on the transmission hump beside his left foot to add drive to the front wheels, and this can be done without any reduction in speed. For very exacting work such as pulling a lorry out of a ditch or climbing really steep gradients (1-in-2 is the limit) the vehicle must be stopped, and the red knobbed lever is pulled backwards, engaging the lower ratio in the transfer gear box. The driver still changes gear and uses the clutch as before, but effective gearing is very much lower. The overall ratio in bottom gear with transfer gearbox in use is 39.7-to-1, and the road speed at 4,000 rpm is only 8 mph. Maximum speed of the four-cylinder in top gear with low ratio selected is 33 mph.

The transfer gearbox automatically brings in four-wheel drive even if it had not been previously selected, so that the enormous torque multiplication is shared by the two final drives. If only four-wheel drive has been used, by depressing the yellow knob, it is disengaged by stopping and pulling the transfer gear lever momentarily back towards the low ratio position and then forward again. Four-wheel drive should not be kept in use on the hard because it affects the steering and accelerates tyre wear by wind up in the transmissions; front and rear wheels inevitably turn at slightly different speeds.

These are the chief features of the Land-Rover, so well known to its thousands of enthusiastic users. We have always felt there was a case for making the interior a little less ugly but we seem to be alone in objecting to it, and it was certainly cleaned up considerably with the changes introduced in April last year. Clearly it is just what owners want, and makes one of Britain's proudest success stories. **S.B.**

# SUCCESS STORY

## 20 YEARS OF LAND-ROVER PRODUCTION

# YOU NAME IT...

## ...AND LAND-ROVER CAN SUPPLY THE VERSION FOR THE JOB

### Identity Parade

The vehicles in the picture are as follows, left to right:

**Back Row:** 1, 88 Basic; 2, 88 Hardtop; 3, 88 with cab and ¾-length canopy; 4, 109 Hardtop; 5, Forward control four-cylinder; 6, Forward control six-cylinder; 7, 88 Diesel; 8, 109 with cab and ¾-length canopy; 9, 109 ten-seater with tropical roof.

**Second Row:** 1, 109 with Simons elevating platform, 25ft lift; 2, 88 with articulated trailer workshop for Electricity Board; 3, 109 with moving conveyor, adjustable height.

**Third Row:** 1, 109 Estate car with mobile cinema (speakers and screen); 2, Shorland armoured car; 3, Forward control with compressor.

**Fourth Row:** 1, MAP winch (5,000lb pull) 109 Estate Car; 2, 109 Hardtop with Broom and Wade compressor and hydraulic drum winch (4,000lb pull); 3, 109 Emergency Ambulance, sometimes used as a hearse for deaths in awkward places; 4, Pilchers Ambulance; 5, Martin Walter four-berth Dormobile on 109 Estate Car; 6, Fire fighting unit on forward control 110.

**Front Row:** 1, 88 Swift forestry vehicle with winch between the seats, 3,500lb pull and 1,000ft of cable; Sawbench trailer; 2, 109 with Atkinson-Howie snow blade, hydraulic height adjustment; 3, 109 Truck with Evers and Wall crop sprayer; 4, Harvey Frost vehicle recovery crane on 109 truck; 5, 109 Fire Fighter; 6, 88 Estate Car with Normalair refrigeration.

*Prices for many of the special bodies are strictly "on application" and depend on the exact specification ordered. They are also varied by the type of chassis chosen, and choice of engine. Here are some of the more interesting special bodies and specialist equipment*

| RECOMMENDED PRICES: | UK List £ s d | Purchase Tax (where applicable) £ s d | Total £ s d |
|---|---|---|---|
| REGULAR—88in. WB—4 cylinder—Petrol | 750 0 0 | — | 750 0 0 |
| REGULAR—88in. WB—4 cylinder—Diesel | 860 0 0 | — | 860 0 0 |
| **109in. WB LONG:** | | | |
| Basic—4 cylinder—Petrol | 860 0 0 | — | 860 0 0 |
| Basic—6 cylinder—Petrol | 925 0 0 | — | 925 0 0 |
| Basic—4 cylinder—Diesel | 970 0 0 | — | 970 0 0 |
| De Luxe—4 cylinder—Petrol | 885 0 0 | — | 885 0 0 |
| De Luxe—6 cylinder—Petrol | 950 0 0 | — | 950 0 0 |
| De Luxe—4 cylinder—Diesel | 995 0 0 | — | 995 0 0 |
| **STATION WAGONS:** | | | |
| 7 Seater—88in. WB Regular—4 cylinder—Petrol | 910 0 0 | 254 17 3 | 1,164 17 3 |
| 7 Seater—88in. WB Regular—4 cylinder—Diesel | 1,020 0 0 | 285 8 4 | 1,305 8 4 |
| 10 Seater—109in. WB Long—4 cylinder—Petrol | 1,060 0 0 | 296 10 7 | 1,356 10 7 |
| 10 Seater—109in. WB Long—6 cylinder—Petrol | 1,125 0 0 | 314 11 8 | 1,439 11 8 |
| 10 Seater—109in. WB Long—4 cylinder—Diesel | 1,170 0 0 | 327 1 8 | 1,497 1 8 |
| 12 Seater—88in. WB—4 cylinder—Petrol | 1,070 0 0 | — | 1,070 0 0 |
| 12 Seater—109in. WB—6 cylinder—Petrol | 1,135 0 0 | — | 1,135 0 0 |
| 12 Seater—109in. WB—4 cylinder—Diesel | 1,180 0 0 | — | 1,180 0 0 |
| **FORWARD CONTROL:** | | | |
| 110in. WB 6 cylinder Petrol, with Platform Rear Body | 1,276 0 0 | — | 1,276 0 0 |
| 110in. WB 6 cylinder Petrol, with Fixed Sides Rear Body | 1,296 0 0 | — | 1,296 0 0 |
| 110in. WB 6 cylinder Petrol, with Drop Sides Rear Body | 1,306 0 0 | — | 1,306 0 0 |
| 110in. WB 4 cylinder Diesel, with Platform Rear Body | 1,340 0 0 | — | 1,340 0 0 |
| 110in. WB 4 cylinder Diesel, with Fixed Sides Rear Body | 1,360 0 0 | — | 1,360 0 0 |
| 110in. WB 4 cylinder Diesel, with Drop Sides Rear Body | 1,370 0 0 | — | 1,370 0 0 |

(WB = Wheelbase)

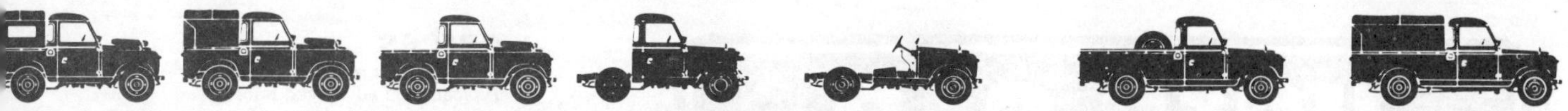

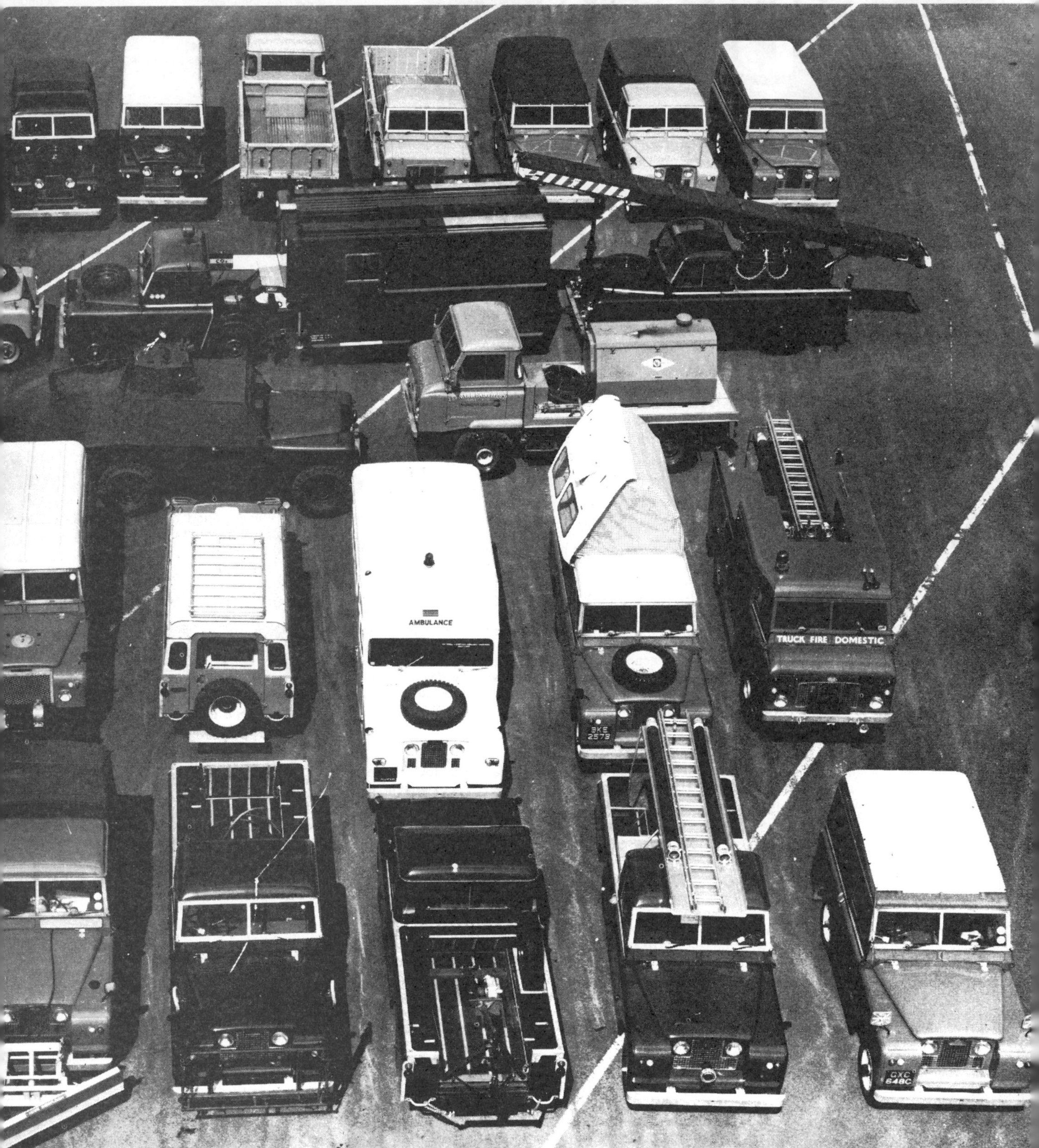
AMBULANCE
TRUCK FIRE DOMESTIC
GXC 648C

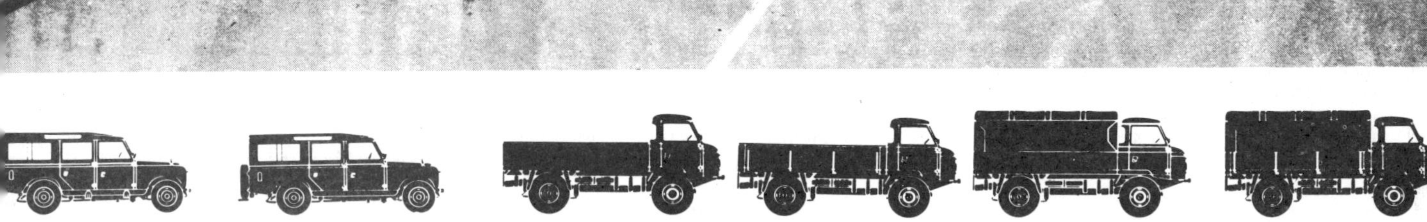

# USED CAR TEST

## 283: 1964 Land-Rover Searle 88

### PERFORMANCE CHECK
(Figures in brackets are those of the original Road Test, published 19 November 1965)

| | | | | | |
|---|---|---|---|---|---|
| 0 to 30 mph | 6.2 sec ( 6.9) | 10 to 30 mph | 11.0 sec (10.2) |
| 0 to 40 mph | 11.7 sec (12.3) | 20 to 40 mph | 11.5 sec (13.8) |
| 0 to 50 mph | 20.8 sec (19.9) | 30 to 50 mph | 15.0 sec (13.2) |
| 0 to 60 mph | 42.7 sec (36.1) | 40 to 60 mph | 29.1 sec (22.8) |
| Standing ¼ mile | 25.3 sec (24.0) | Standing km | 49.0 sec (——) |

In top gear:

## PRICES
| | |
|---|---|
| Car for sale at Marylebone at | £565 |
| Typical trade advertised price for same age and model in average condition | £550 |
| Total cost of car when new (free from tax) | £902 |
| Depreciation over 4 years | £337 |
| Annual depreciation as proportion of cost new | 9.3 per cent |

## DATA
| | |
|---|---|
| Date first registered | 19 June 1964 |
| Number of owners | 1 |
| Tax expires | 31 October 1968 |
| MoT expires | 17 July 1969 |
| Fuel consumption | 16-20 mpg |
| Oil consumption | 500 m.p.pint |
| Mileometer reading | 34,915 |

## TYRES
Size: 6.00—16 in. Avon Traction on all wheels. Approx. cost per replacement cover £10 0s 6d. Depth of original tread 12 mm; remaining tread depth 6mm (left, front); 5mm on each of other three; just over legal minimum on spare.

## TOOLS
Jack, wheel nut spanner and starting handle. No handbook.

## CAR FOR SALE AT:
Archie Simons Ltd., 8-20 Gateforth Street, Marylebone, London, NW8. Telephone: 01-262 9530.

---

NOW that another chunk of purchase tax has been added to the cost of new cars, Land-Rovers seem even better value because in simple hardtop or canvas tilt form they are tax free. It is therefore against the cost of a new one that the used car prices of Land-Rovers should be assessed, and this four-year-old example is an indication of how well they keep their values. On the £750 price of the cheapest new one available, the saving is only £185. Perhaps this also confirms how well such a rugged vehicle lasts; indeed with its strong chassis and all-aluminium bodywork the Land-Rover is almost ever-lasting.

In this case the prices are slightly distorted by small allowance for the price of the Searle Safari sleeper conversion, one of the simpler of the camping conversions available from R. J. Searle Ltd., of Sunbury. At the rear there is an ingenious pull-out framework for a tent extension, and the tent itself is with the vehicle, complete with groundsheet, but there are no pegs or mallet. The beading on the tent edge is now rather frayed and is very difficult to thread round the slot at the rear of the vehicle. A table is fitted, and there are bench seat cushions on one side which can be converted into a bed. There are none of the cooking, washing and general camping facilities normal in a fully-equipped motor caravan.

The log book reveals that a new engine has been fitted; and in all mechanical respects the vehicle has either lasted extremely well or has been very well maintained. There is not much difference to be detected, when driving, between this and an almost new Land-Rover. Slight vibration occurs occasionally on closing the throttle abruptly at low speed, suggesting wear somewhere in the transmission, and the brakes tend to squeal when the pedal is pressed hard; but there are no other faults.

Prompt starting is customary with Land-Rovers, and this one is usually away at the first touch on the button, hidden away beneath the central instrument panel. Choke can be pushed in almost as soon as the engine has started and there is no stalling. Low gearing and a lot of torque at low revs make it a very tractable vehicle, able to accelerate smoothly even from as low as 10 mph in top gear. There is not much need for high revs, and only if gear changes are needlessly delayed is there some roar and vibration.

Land-Rovers are best driven more as lorries, with upward and downward gear changes always made in good time. Clutch take-up is very smooth, and although the almost vertical action of the pedal accentuates the leg effort needed, it is not too heavy. The gear change is positive and very easy to use, provided changes are not hurried, otherwise it is easy to crash the non-synchromesh first and second gears. Double declutching helps and is important when changing down.

Vague steering is the Land-Rover's weak point, perhaps accentuated by wear on this used example. The control is designed to eliminate feed-back on rough going, enabling the driver to keep control easily when pounding over the rough, but on ordinary roads continual wander has to be checked because of the general steering looseness. The springs are also designed for rough going and are very hard for riding unladen. On moderate surfaces the ride is not too bad, but on potholes and undulations not a bump goes by without transmitting lively reaction, and the joggy, bouncy ride on secondary roads is tiring.

Very good braking is available, and pedal loads are not excessive. The handbrake is also extremely efficient, and works on the rear transmission; severe vibration results if any attempt is made to apply a Land-Rover handbrake while the vehicle is still moving.

Buyers are sometimes tempted by the ruggedness of a Land-Rover, but it can be a mistake to buy one if there is no need for the cross-country ability. In traffic, they run away with a lot of fuel and we obtained only 16 mpg on the road with some London traffic driving included. But all these disadvantages are forgotten in the cross-country conditions for which it is intended. The four-wheel drive and low ratio transfer gearbox controls are working correctly.

Archie Simons Ltd. specialize in used Land-Rovers and this is typical of their stock of about 80. It is very sound mechanically, and apart from some detail matters is in reasonably good body condition as well.

## Condition Summary

### Bodywork
It is not expected that a Land-Rover will be kept in pristine condition, and the utilitarian finish inside is intended not to show wear. Everything is well soiled, but condition is quite reasonable, and the simple transverse seats are fairly comfortable. There is an ugly pole for the tent structure, for which there is no carrying provision to allow it to be stowed out of the way, and the tent itself is a large, cumbersome item. The bodywork is sound, and although the pale blue finish has faded and become very dull, there is no corrosion, the panelwork being all in aluminium. Underneath, the condition is good except for the exhaust system which is extensively rusted. Almost total lack of chromium plate eliminates other sources of visible deterioration.

### Equipment
Lighting equipment is a possible weak point on the Land-Rover, as water finds its way inside the lamp units. One of the side lamps and the indicators on the left were not working for this reason. All other equipment is working correctly, including the three instruments—speedometer, ammeter and fuel gauge, horn and wipers.

### Accessories
The most expensive item on the vehicle is a tropical roof, which certainly keeps the interior cool in hot weather. An additional windscreen wiper for the passenger has been added and there is a recirculating heater. Britax lap and diagonal seat belts are fitted, but the buckle on the passenger's side is not working because the spring has come adrift. There are also a reversing lamp, foglamp and two wing mirrors. A large roof rack is fitted at the back.

*Above: The Searle Sleeper Conversion includes this tent annexe with pull-out frame. The Land-Rover bodywork has lasted well although the paintwork has faded. There is quite a lot of rust on the engine, yet it is in very fit condition. Left: Not surprisingly the interior is well soiled but there is no structural deterioration*

## About the Land-Rover

THE first Land-Rovers were produced in July 1948. There have been many revisions to engines and chassis lengths in the 20 years' production, but the basic engineering is still largely the same. The original vehicle had an 80in. wheelbase, with a 1,595 c.c., 50 bhp engine, derived from that of the Rover 60 car. Four-wheel drive with two speed ranges was standard, and there was a freewheel. Only an open model was available at first, but an estate car was released in autumn 1948 and a plain hard top in 1950; the freewheel was deleted in 1950. In 1952 a 1,977 c.c. engine was introduced, this being a bored out and revised version of the original 1.6 litre.

In 1954, the 80in. model was replaced by an 86in. wheelbase Land-Rover, and later in the year a long-wheelbase (107in.) version followed; this latter model had a payload of ¾-ton. Only two years later, 88in. and 109in. versions replaced the 86 and the 107in.; re-circulating ball steering and foot-dipping were introduced. Just a year later came the first diesel option—not a conversion from any of the petrol engines—with 52 bhp at 3,500 rpm, and a capacity of 2,052 c.c.

Series 2 Land-Rovers were announced in Feb. 1958, having slight styling changes and a new optional 2,286 c.c. petrol engine. This was based on the sturdy diesel option, producing 70 bhp at 4,250 rpm. The 107in. Series 1 station wagon continued until September, to be replaced by the 109in. station wagon in 1959.

All models became Series 2A in September 1961, and the optional diesel engine was enlarged (by boring) to 2,286 c.c. to commonise it with the petrol unit. Power output became 68 bhp at 4,250 rpm.

One year later came the only major change in Land-Rover shape with the introduction of the 109in. forward control version. The cab was, in effect, moved forward to a position above the front wheels and straddling the engine; wheels were now 9.00—16in. to deal with the extra payload. It took another four years before there were more changes. In September 1966, the 110in. forward control version became available with a 6-cyl., 2,625 c.c. petrol engine developing 85 bhp at 4,500 rpm, a less powerful version of a car unit that had been in production for some years. To improve stability, the track was widened by four inches, a front anti-roll bar added, and rear springs re-positioned.

This 6-cyl. engine became optional in 109in. normal control Land-Rovers from April last year, as a further alternative to the 2.3-litre petrol and diesel units. Normal control versions were then given much revised instrumentation and controls, a more convenient handbrake and a concealed wiper motor. Later the company announced more comfortable optional front seats, the first attempt to suit the Land-Rover to changing markets.

All three versions—88in., 109in. normal control, and 110in. forward control—remain in full production.

# SANDTREKKER

## Six-wheel-drive Land-Rover
By J. C. Baker

**Purchased for £220, a short wheelbase Land-Rover was converted to long chassis and two extra driven wheels added. It then served as transport for four from the Midlands through Cairo to South Africa.**

THREE of us (Tony, Roy and myself) were looking for a vehicle to undertake the 14,000-mile overland journey to South Africa. As we were short of money and could not hope to afford a new Land-Rover, we optimistically attended the army disposal auction at Ruddington in the hope of finding a cheap second-hand one. It was the very last lot of the day and we were beginning to lose heart. We had only £220 and even the old series I Land-Rovers were fetching £300. The hammer fell at just £220. It was our bid, the last of our money and we had bought it "unseen".

The first view of Lot 360 was not reassuring; a battered, clutchless, two-wheel-drive Land-Rover, standing up to its axles in mud in a corner of the field. It had been used by the army as a training vehicle and L-plates were still painted on the front and back.

It took six months to rebuild and then it was to all intents and purposes a conventional 1965 88 in. Land-Rover station wagon which we proudly took for its first trials in the Scottish snow. It was on this holiday that the complication arose; we recruited another crew member. Graham was certainly going to be an asset to the expedition but now our 88 in. station wagon was not going to be big enough to carry all four people and their luggage, together with the extra water and fuel tanks which we anticipated would be necessary.

The obvious solution would have been a 109 in. model but financially this was out of the question. A trailer seemed to be a possible answer until someone, in jest, suggested lengthening our own chassis and propping up the extension with an extra axle. We looked at each other, waiting for someone to laugh. Nobody did. The Sandtrekker idea was born.

We made drawings and submitted the ideas in principle to Rover's engineers. They were discouraging and pointed out the problems of tyre scrub at the four back wheels and unmanageable steering. The unit would be grossly underpowered and a real danger existed in using an unstressed chassis. In fact they strongly recommended that we forget the idea and start saving for a real long-chassis Land-Rover. With the sublime self-confidence of ignorance, we decided to start work at once. Our self-imposed time allocation was so short that every day was going to be needed.

The rear body was removed and the chassis lengthened by cutting a short section out of the rear end of the frame and replacing it with a similar but longer section. The original Land-Rover strength and stability were restored by the addition of various stiffening members and crossbraces as shown in sketch 1. To eliminate some of the buffeting the new rear suspension would inevitably receive, a system was designed on basic close-coupled axle principles whereby the two leaf springs were connected by a bell cranked lever, in a way which meant that only half the wheel movement was transferred to the body.

The new body was constructed by cutting the back end from our station wagon and the front end from a scrap body. The two halves were riveted together. Now we really had room to spread ourselves: 84 in. of it compared with only 43 in. in our original little Land-Rover. To give protection against the tropical sun we sprayed the panels white. The tropical roof was Day-Glo red, testifying to our simple faith in the efficiency of the RAF's desert rescue service. The matt black bonnet was perhaps a mistake—but it looked good!

Five weeks after we had started we were able to drive our six-wheel Land-Rover into the Rover factory at Solihull, once more to seek the advice of the experts; once more our complacency was sadly bruised. After the trip round the test track, amid much shaking of heads, doubt was expressed about the braking and steering. They also considered that the four-wheel drive, operating as it did only on the front and middle wheels, would produce insufficient traction for a vehicle of Sandtrekker's size working in severe conditions. They suggested that we forget the whole thing but added that if we wished to proceed with our ridiculous idea then we could have the use of one of Rover's fully equipped workshops and any assistance they could offer. We were speechless!

In Cornwall that summer, a dramatic six-wheel skid underlined their warnings that the springing and braking systems were wholly inadequate. As these are highly specialized fields and we had neither the necessary knowledge nor sufficient capital to solve these problems satisfactorily ourselves, we approached the leading manufacturers for help.

Bramber Engineering took up our springing problems with tremendous enthusiasm and even went so far as to send one of their technical advisers all the way from South Wales to discuss the problem. As a result of this meeting, Bramber produced a set of springs that suited our conditions exactly and withstood every shock of the 14,000-mile journey. Our braking problems were efficiently solved by Girling who studied our particular conditions as though we were a large firm contemplating a big order. They even loaned us a technician for a few days to install all their fittings to their own high standards. These included a Clayton Dewandre servo unit which we managed to install in the engine compartment only with great difficulty. After the installation of the Girling-Clayton Dewandre system Sandtrekker could stop on a sixpence!

To combat lack of traction, an idea emerged for powering the rear axle. The Land-Rover transfer box consists basically of a chain of three cluster gears, the last of which provides power by means of a selector to either the rear or front and rear axles. It was our plan to introduce a fourth gear into the power chain which would drive a further prop shaft (sketch 2). So, back to the drawing board and eventually a design was produced which worked in theory. Parts were made (in the most unlikely

Useful "cow-catcher" at the front carried part of the extra petrol and the water tanks
Sketch 1: The extended Land-Rover chassis with some of its modifications

# SIX-WHEEL-DRIVE LAND-ROVER

places) and assembled with our existing gearbox. Now a power line was required to transfer the power from the gearbox to the third axle. It would have to pass over the centre axle and bolt to the rear differential and have enough spline movement to allow both axles to be in any stage of change between full bump and full rebound. With increasing effrontery we approached the experts, Hardy-Spicer, who readily provided us with the answer in the form of three shafts, the middle one being fixed to the centre axle casing and moving up and down with it. The angles of operation were far in excess of what Hardy-Spicer normally recommended but we had so many untried systems on board already (including the crew) that we accepted the risk.

### Plenty of fuel

Provision was made for 58 gallons of petrol. Four $4\frac{1}{2}$ gallon jerricans were carried behind a heavy cowcatcher built over the front bumper, four jerricans were carried on the roof rack and an additional tank was fitted under the front passenger seat; 30 gallons of water were carried in two tanks.

By this time, having lost all sense of proportion, we coated the floors with thick Wilton carpet, built specially contoured seats for driver and navigator (though the rear passengers made do with seats from an old Britannia) and pampered ourselves with luxury trim and separate heaters for front and rear. The generosity of Smiths Instruments gave the facia almost aircraft appearance.

On the morning of 3 March, 1967, we put Sandtrekker into gear for the first time under power, and set out three days later for Port Elizabeth, South Africa.

To our surprise, everything worked perfectly. It seemed that we might reach Africa after all. The first dramatic episode occurred in Europe while climbing in the Pyrenees. The altimeter was showing 12,000 ft and the weather took a turn for the worse. Driving snow necessitated the use of the heated screen and the wipers were set to full speed in an effort to see through the swirling mists. High above us a noise like thunder brought us to a sharp stop and a few seconds later the road in front was struck by a falling mass of snow and ice. There was a mile long queue of traffic waiting for the snow plough when we finally decided to climb over the obstruction in six-wheel drive.

### Customs delays

Most of the main roads from Tangier to Egypt have been surfaced since Rommel's day and although today many stretches are in bad need of repair, the innumerable forms at the many customs posts provide more headaches for the North African traveller.

For us, this part of the journey was both colourful and interesting. It also provided an opportunity to test Sandtrekker in the sand from which we named her. These tests were successful beyond our wildest hopes and so it was with confidence that we moved on to Egypt. Cairo proved to be everything we had expected of it; dirty, romantic, disorganized and intriguing, a bewildering pattern of sight and sound.

We managed to leave Egypt and the Sudan

*Nearly, but not quite, stuck in the African mud*

*This is a road!*

*This side view shows how long a six-wheel-drive Land-Rover becomes*

*Smiths did the vehicle proud with extra instruments*

*Rehearsal for "rigging" the land ship*

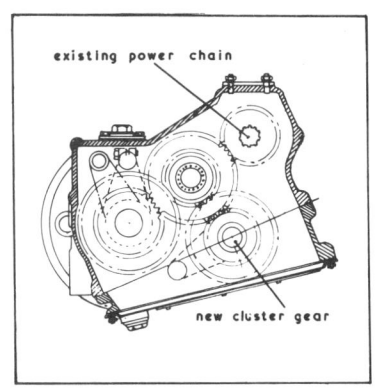

*Sketch 2: Modifying the power chain to drive an extra propshaft*

before war with Israel was declared. We had to leave the Sudan by sea as the frontier was closed and in the Yemen we were kept prisoners in our cabins. Our guard was armed (rather impractically) with a bazooka. Fortunately, the ship was Yugoslavian and so it was allowed to proceed.

The British Ambassador in Addis Ababa strongly warned us against continuing south. Not only would the imminent rains make the roads impassable but the Shifta bandits were laying mines on all the roads into Kenya. With a war raging behind us there was no alternative but to go on and in Addis at least we were safe for a while. Mobil Oil were our hosts and gave us the time of our lives. For a fortnight all our problems were forgotten.

The rains started just outside Dilla. The road was incredibly bad, huge boulders and great potholes with the occasional 2 ft crevice crossing the road from side to side. After about a mile we met our first real mud. In places it was feet deep and on either side of the track it was as high as the vehicle. In low range six-wheel drive Sandtrekker found grip where it looked impossible and, after two minutes, we were through. We were all thrilled by the way our vehicle performed and felt we could tackle anything, but after 200 miles of it we were not so sure. They took us four exhausting days. We could generally pick the stretches where the mud was only axle deep but we sometimes made mistakes. We also found we could hire a whole African village to dig us out for just 12s 8d!

After Yebellow the mud diminished as we moved under the influence of the East African climate.

**Taking a risk**

During the previous week 10 people had died trying to get through from Mega to Moyale but the day we arrived in Mega an army convoy had successfully covered 75 miles in the opposite direction. In theory all we had to do was travel in their tyre tracks. The driver of a big Beryl five-tonner was willing to make the run so, rather than hang around for three weeks waiting for the army, we decided to join him. The local governor tried to persuade us to wait until the situation had improved but we decided to push on because the longer we waited, the greater was the chance that the Shifta would lay a new set of mines, this time in the freshly made tracks of the army convoy.

We used most of our camping gear to pack around our legs in an attempt to lessen the explosion, should it come, then reluctantly drove off in the wake of the brave lorry driver.

The going was slow and the path difficult to follow; the road was unbelievably bad. Great crevasses split its surface, there were again pockets of mud and steep ridges of rock suddenly appeared out of deep sandy holes.

After an hour of travelling we came across a wrecked Toyota in which a man and woman had lost their lives. Some miles farther on the remains of a tyre and a few fragments of twisted metal scattered round a deep crater were all that remained of a Land-Rover in which eight people had been travelling. The hours dragged painfully past until we finally drew into Moyale, miraculously intact.

We took 200 days after leaving England to reach Port Elizabeth and everywhere we travelled Sandtrekker aroused considerable interest. She was even tested by the Armed Forces of both South Africa and Rhodesia. We were impressed by the friendliness and open handed generosity of so many of the people we met and this made our time in Africa most enjoyable. Now only one relic remains of our memorable experience . . . does anyone want to buy a Land-Rover? □

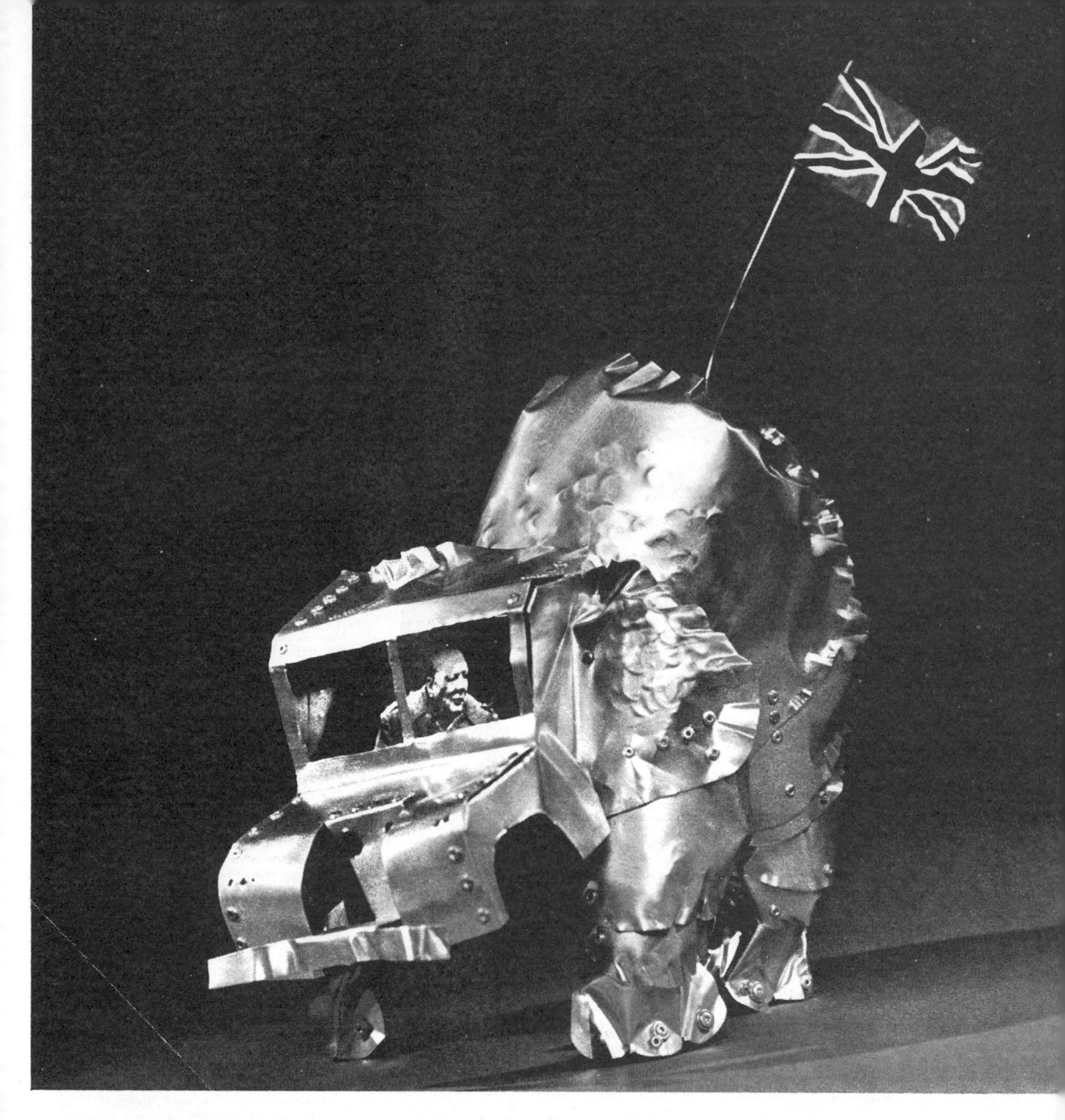

# O'KANE ON LAND ROVERS

### BY DICK O'KANE

ON THAT GREAT bright day when I manage to accumulate the time, money and space all at once, I'm going to go out and buy a TC, a D-Jag, an XK-120M (which I'm going to bolt a C head onto and call an "XK-120MC" so you can all write righteous letters to the editor), a Mark IV (yes, I know) which I'm just going to look at and never drive, and a Land Rover.

Now, I'm fully aware that I don't need any of that stuff, least of all the Land Rover. But I'm going to drive the Land Rover. A lot. Simply because Land Rovers turn me on.

I got into the Land Rover thing about ten years ago when trying to sell business machines to Mom-and-Pop grocery stores in an economic disaster area finally got to my soul and my bank balance simultaneously. So I wound up in a very small, very informal Land Rover agency as a salesman, sometime mechanic, chief gopher and seat warmer.

The first thing you learn about a Land Rover is that it has a personality which is uniquely its own. To me, a Land Rover is a safe, warm, comfortable place to be. When in a Land Rover you're safe from any assault by man or nature.

CONTINUED ON PAGE 100

# Twenty-one this year

*A quick look back on the Land-Rover Saga.*

### By Harold Hastings

WHEN I was first shown a Land-Rover some months before it was revealed to the public in 1948, I was full of admiration tinged with a slight doubt. What worried me was whether it might not be a little *too* good! Rover engineers had been concentrating on quality and refinement since the Wilks brothers (Mr. S. B. Wilks and the late Mr. Maurice Wilks) took control in 1933, and the company had, moreover, been building and developing jet engines during the war. The whole set-up was geared for Quality with a capital "Q", whereas it seemed to me that a little agricultural crudity might have served equally well for a cross-country vehicle of this kind, and would have provided substantial economies in production.

How wrong can you be! What Rover had produced was, in fact, just what the world had been waiting for and now, over 21 years later, it is going more strongly than ever, in spite of strong competition. At one time, its competitors in various parts of the world numbered over a score and are still not far short of that total; but there is still only one Land-Rover—a name, incidentally, which was as clever a conception as all the rest since it incorporates its make name and purpose in three syllables. Even the august BBC in the days when proprietary names were strictly taboo over the air, could not avoid using it until they finally thought up the term "field car".

Looked at now with hindsight, I think that it was probably the care and refinement in engineering detail, almost as much as the soundness of the general conception, that made it such a success, because the car-like quality widened its scope enormously.

Perhaps the oddest thing about the Land-Rover is that it owed its origin to a quirk of circumstance. Britain was in desperate need of foreign currency in the years immediately after the war (and has been ever since) and the situation was such in 1946 that the Government of the day took the drastic step of rationing steel in accordance with the export potential of individual car makers.

To a manufacturer whose activities were centred in the production of luxury cars with a very limited export potential, this was something of a facer. Sheer survival demanded something that would sell in worthwhile numbers abroad as well as at home. To the peaceful and non-industrial atmosphere of Anglesey, the brothers Wilks retired to think out the problem away from the day-to-day cares and frustrations of the factory; and it was here that the Land-Rover was born.

This was early in 1947 and time was short. Only a few months later prototypes were running about—not only over a vicious cross-country circuit in the grounds of the factory and up and down the 45-degree slopes of the works air-raid shelters (which was where I first tried one), but were also to be seen doing all sorts of odd jobs like pulling hay-raking equipment, dragging an 8 ft. harrow and driving things like threshing machines and circular saws from a power take-off. The outcome of all this was revealed to the public little over a year after the Anglesey conferences when the car was exhibited for the first time at the Amsterdam Motor Show which opened on April 30, 1948. What happened after that is public knowledge, although a little recapping is appropriate here.

In its first two years of production, 24,000 Land-Rovers were produced, with an earning in foreign currency of no less than £5-million. By 1959, the quarter-millionth Land-Rover had come off the lines, and by the beginning of April 1966, total production had reached the half-million with an overall earning of £230-million in foreign exchange.

To bring the story up to date, total production at the time of the Land Rover's 21st birthday in April this year, stood at 644,000 units (of which over 70% went to more than 170 overseas markets) and models are now rolling off the lines at well above 1,000 a week. If present production continues until the end of the year, output will reach its best-ever annual total of more than 50,000.

Throughout this period, the company has never had to depart from its original conception although progress in design has been continuous and the number of variations on the original theme vastly increased—as will be gathered from the landmarks in development outlined in the accompanying panel. As for the range of adaptations and accessories to enable the Land-Rover to fulfil ever-widening functions in agriculture and industry, these, too, are greater than ever before.

---

## Land-Rover Landmarks

**1948.** Land-Rover introduced in April, with 80-in. wheelbase and 1.6-litre P3 petrol engine (as in Rover 60 car). Seven-seater station-wagon version added as alternative to "regular" model later in year.

**1950.** Plain hard-top version produced. Transmission modified by eliminating free-wheel in front drive and arranging for high range of gears to operate on rear wheels only unless front drive specifically selected.

**1952.** Original 1.6-litre engine superseded by 2-litre engine to improve low-speed torque.

**1954.** Wheelbase of existing model increased to 86 in. and long wheelbase (107 in.) model introduced with ¾-ton payload.

**1956.** Wheelbase of 86-in. model increased to 88 in. and 109-in. wheelbase model produced to accommodate diesel engine (introduced in following year).

**1957.** 2-litre diesel engine offered as alternative to petrol.

**1958.** Series II introduced with improved bodywork and new 2¼-litre, 4-cyl. petrol engine with full o.h.v. (in place of o.h. inlet and side exhaust). Diesel engine unchanged. Production of 107-in. station wagon discontinued in September.

**1961.** 2-litre diesel engine replaced by 2¼-litre unit. All models Series IIA.

**1962.** 12-seater station wagon and new forward-control model introduced on 109-in. wheelbase.

**1966.** Introduction of 110-in. model offered with six-cyl. 2.6-litre petrol engine, 2¼-litre petrol engine (overseas only) or diesel engine; track increased by 4 in., anti-roll bar added at front, and stability improved at rear by wider spring base.

**1967.** Six-cylinder petrol engine made available (as alternative) in 109-in. model. Many detail refinements to controls and equipment incorporated.

**1968.** Headlamps moved from grille to wings for certain overseas markets to comply with local laws. Later in year, 1-ton version of 109-in. model introduced.

**1969.** Headlamps moved into wings on all models.

*Hamish Cardno treads a forgotten world*

# up and over

*Storming up Corrieyairick Pass by Land Rover*

*The foundations of the old road are still in evidence but the lack of a top dressing makes a high ground clearance essential.*

PICTURES BY PAUL SKILLETER

HIGH UP in the mountains which run through the centre of Scotland runs a road—a by-pass or back-double, to debase it—which provides a direct link from the main Perth-Inverness road, the A9, on one side, to the shores of Loch Ness on the other. It's the Corrieyairick Pass.

In places it is narrow but well-surfaced, in others it is an uneven pathway of boulders and smaller stones rather like the bed of a dried-up stream; in others again it is a series of serried ridges through marshy ground, and for a short stretch it is like a well-made Forestry Commission road.

At one time it was the highest through road in Britain, climbing over desolate and breathtakingly lovely country from Laggan Bridge to some 2,500 ft. above sea level on the slopes of the Monadliath Mountains, then stumbling down through its foothills to emerge at Fort Augustus, at the southern end of Loch Ness.

It is no longer a through road—at least not to any normal vehicle. The ubiquitous General Wade built it as a pass through to the Western Highlands in 1732 when he was despatched north of the Border to tame the "North British" who were at that time rebellious, to say the least.

## up and over

Wade realised that the only way the Highlands could be controlled was by having proper roads so that his troops could quickly get from their garrisons to the various trouble spots. If history does not rate him along with Mark Anthony and Nelson in the ranks of the outstanding fighting men of their time, he does have more permanent memorials in that many of the main roads in Scotland to this day are based on the lines through the hills chosen by him as Military Roads "for the better communication of his Majesty's troops"

The Corrieyairack Pass is not one of those roads which was maintained until road building became the science it is now, and there are no records of it being maintained since the early 19th century. Unlike its nearest rival for sheer altitude and loneliness—the Lecht Road from Speyside to Deeside—it suffers from the presence of an alternative route round the eastern shore of Loch Laggan which is less of a civil engineer's dream or nightmare, and which offers more conventionally beautiful scenery.

The Corrieyairack is beautiful not only for the unparalleled views of Loch Ness and the adjoining mountain ranges which can be seen from it, but by virtue of its sheer remoteness. Only the damnable presence of electricity lines and pylons which traverse it bring one crashing back into the 20th century.

We learned of the past—and subsequently of something of its history—from Norman Milne, a Scot who is public relations officer for the Automotive Products Group, and the world's leading authority on the road which he has crossed himself several times, both by foot and on four wheels.

Four wheels were going to take us across (nobody on our staff is a very enthusiastic hill-walker), and it was four wheels which could all be driven at the same time which were necessary, we were told. A Land-Rover was the obvious choice, and after a lot of thought and discussion with experts, we had a short (88 in.) wheelbase model with the latest 2.6-litre petrol engine—specially fitted with an engine-driven winch supplied by Mayflower Automotive Products.

So with our marigold Land-Rover and Mayflower winch we blossomed forth, from the Fort Augustus side of the Pass.

It wasn't very difficult at first, just rocky. The Land-Rover with its big wheels, low gearing and bags of low-down torque just picked its way over the worst of the bumps,

*One of the bridges General Wade didn't build; this one (top left) was constructed fairly recently by the Royal Engineers. Some of the gradients (top centre) leave you standing on your nose— gradients that perhaps accounted for this macabre evidence of the Corrieyairick's isolation (top right). Returning to the northern end, you get some magnificent views of Loch Ness stretching away into the distance (above left), though there is no shortage of water on the pass itself (above right).*

while all the driver had to do was keep a very light foot on the accelerator and steer round the worst boulders.

Then the track smoothed out a bit, and we were able to run along in second or third gear quite comfortably at about 15 m.p.h., slowing only for the larger ruts and stones. It was about this time that we started to appreciate some of the qualities of the Land-Rover a bit more.

In the intermediate, high ratio gears, it is possible to bomb along on unmade roads quite as fast as you want to, but the big brakes are powerful (if heavy) and the machine can be slowed even on very poor surfaces very quickly. You then look at the obstacle which made you slow and decide which gear to tackle it with. During the whole of the crossing we didn't once find an obstacle which the Land-Rover looked at twice. On one or two occasions we were down to low ratio first, which is so low that when you lift off the vehicle all but stops. Slowly but surely the Land-Rover crossed every type of surface—even a water-logged, rutted and treacherous section in a series of zig-zags which raise the road nearly 1,500 feet in a very short distance near the summit—without once giving us cause for alarm, or once having to think about using the winch.

This performance caused the only dissension among the party—an argument (still going on) as to whether a vehicle other than a Land-Rover (or similar four-wheel-drive cross-country vehicle) could make the crossing. One half of the editorial reckons that it's strictly for Land-Rovers, the other that a car with very high ground clearance, very strong suspension, and low gearing to give a reasonable "crawling" speed *could* just do it.

Another pleasant surprise the Land-Rover gave us was its comfort. Ours appeared to be a very de luxe model, with seats which were very well padded and contoured, and there was further padding material under the head lining and in various other places where you might conceivably do yourself an injury.

It was also a quiet vehicle—except on main roads when, as the speed rose, so did the whine from the Dunlop Trakgrip tyres and from the transmission; these drowned any noise there was from the very smooth engine. And we would forgive the tyres anything for the way they crossed what was virtually 14 miles of sharp stones and gi-normous boulders without puncturing, splitting their sidewalls, or letting us (or themselves) down in any way.

As an ordinary road-going vehicle—and we drove it back from Fort Augustus to London just to see how it performed in that respect—it has its drawbacks. Noise, low overall gearing, and a tendency to rattle and shake as you approach the legal maximum speed on motorways, are minus points in this respect, but it was not a vehicle which was ever intended for driving half the length of Britain non-stop. If you want to drive 100 miles up a main road, then short-cut across the mountains, it's ideal.

**M**

# World Vehicle

## Land-Rover Success Story

By Stuart Bladon

This year the Land-Rover celebrated its 21st anniversary; first introduction was at the Amsterdam Motor Show on 30 April, 1948. In subsequent years the consistent problem has always been "we cannot make enough of them to meet the demand". Here we take a quick look at its 21-year history.

ROVER'S original decision to build a working type of vehicle to attract agricultural and industrial markets all over the world was made in 1947, because of the Government's steel rationing at that time. A vehicle had to have high export potential to enable the manufacturer to get the steel he required.

Quantity production started in July of the following year and the original power unit was the 1.6-litre four-cylinder engine used in the Rover 60 saloon, with overhead inlet and side exhaust valves. Many other mechanical components common to the Rover 60 were used. On this first model with 80in. chassis, a canvas hood body was used and later, in 1948, an estate car with seats for seven was added. The British Government's first order for Land-Rovers was placed the following year, and orders began to pour in from overseas. By 1950, in the first two years of its life, the Land-Rover had earned £5 million in foreign currency and 24,000 had been built.

In the same year a hardtop version was introduced. The design was also altered slightly. Originally, drive had been to all four wheels with a free wheel at the front. This was abolished and normal drive

*Wherever roads deteriorate into mud, hills, forest, or tracks through snow, the Land-Rover is the world's choice for tackling them*

came to the rear wheels only with front-wheel drive being added either on selection or automatically when low ratio of the transfer gearbox was engaged. In 1952, a 2-litre, four-cylinder engine was introduced and two years later the wheelbase was extended from 80 to 86in. and a long wheelbase model added (107 in.). In 1956, the wheelbase was extended again to 88in. and the 107in. long wheelbase became 109in. In 1957, a 2-litre diesel engine of Rover's own design was offered as an option.

Series 2 Land-Rovers were introduced in February 1958, with a 2¼-litre four-cylinder overhead valve engine and a 2¼-litre diesel replaced the original 2-litre diesel in 1961. Production passed the quarter-million mark in 1959. The forward control model was introduced in 1962, and at the same time a 12-seater estate car became available.

By April 1966, production had passed the half-million mark and it had earned more than £230 million in foreign currency.

A 110in. forward control model was introduced in September 1966, using the six-cylinder 2.6-litre petrol engine derived from the Rover 100 saloon; and the track was increased by 4in. This six-cylinder engine also became available in the 109-in. long wheelbase range in April the following year as an alternative to the four-cylinder 2¼-litre petrol unit and, a little of the crudity of the interior, which perhaps had contributed to the success of the Land-Rover—by making it look utilitarian and rugged, as well as being so—was taken away by a general tidying-up and re-styling of the interior.

Three basic models continue in production to-day, but there are an enormous variety of modifications, adaptions and special equipment available. Land-Rovers are used by police and military forces of 70 countries and over 70 per cent of production has been exported to approximately 175 different countries throughout the world.

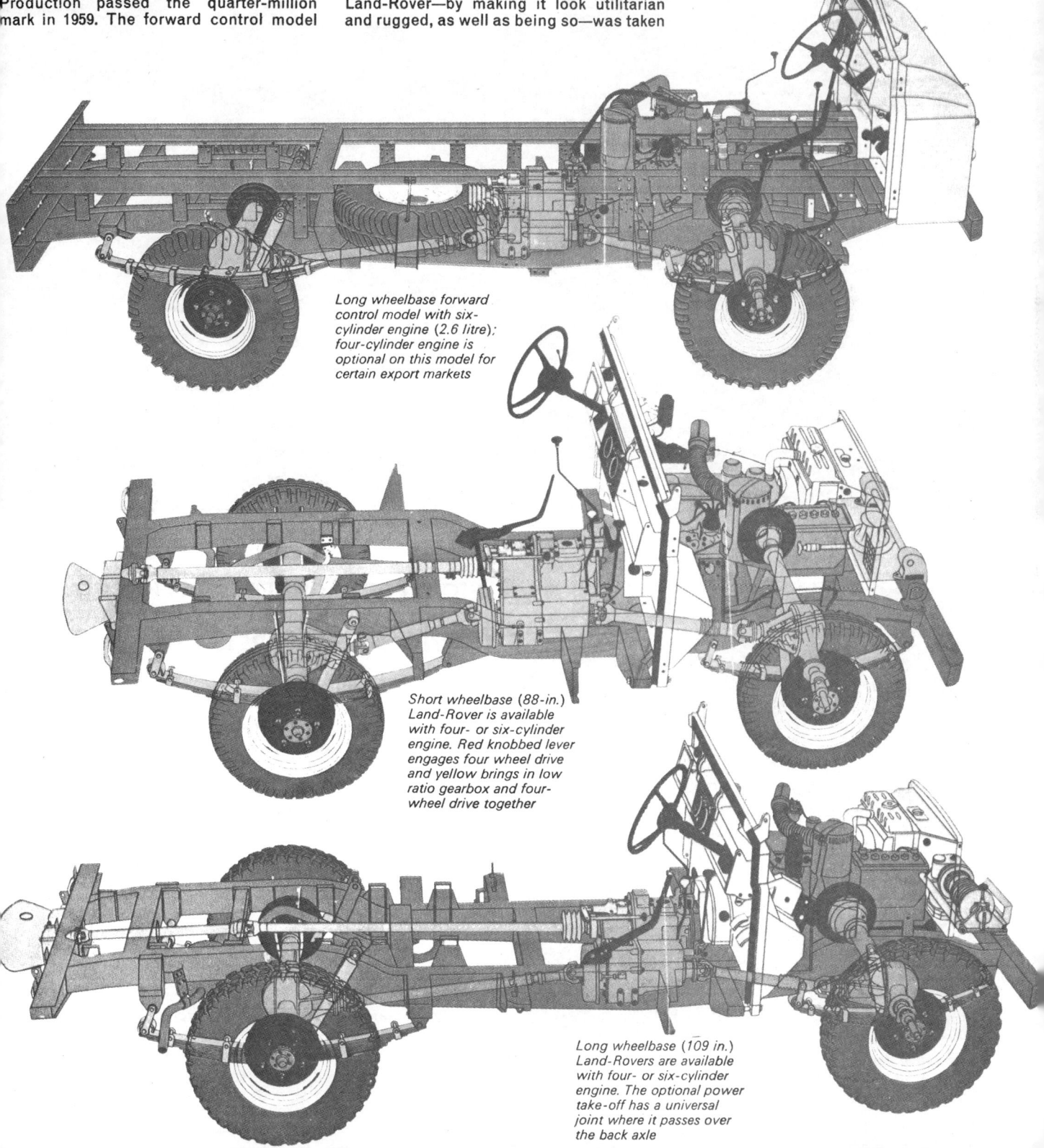

*Long wheelbase forward control model with six-cylinder engine (2.6 litre); four-cylinder engine is optional on this model for certain export markets*

*Short wheelbase (88-in.) Land-Rover is available with four- or six-cylinder engine. Red knobbed lever engages four wheel drive and yellow brings in low ratio gearbox and four-wheel drive together*

*Long wheelbase (109 in.) Land-Rovers are available with four- or six-cylinder engine. The optional power take-off has a universal joint where it passes over the back axle*

# LAND ROVER
## 88 INCHES OF *GET THERE*

*The reach for the 4WD controls is rather long, instruments set too far to the right, but comfort and legroom is great.*

**EIGHT FORWARD SPEEDS, 77 HP ENGINE, AS TOUGH AS A PIECE OF GRANITE—THE NAME "LAND ROVER" IS APTLY CHOSEN**

**Text and Photographs
by BARRY LOPEZ and STAN BETTIS**

We "roved" in this British Import for five days in Oregon's deserts, mountains and rivers and were hard put to find anything to stop it. We drove it across swollen streams, up and down abandoned logging roads, fifty miles across lava strewn deserts, finally coming to a halt against a rock wall 600 feet high. Given a little time we probably would have gotten over that too. The fact of the matter is it was us, not the vehicle, that ran short on endurance.

The Land-Rover is a vehicle with a language all its own. In Roverese "road closed" means "road open," "detour" means "keep going," and "I don't think there're any roads up there" means you may have to put it in reverse a few times.

This go-anywhere, do-anything vehicle does have its drawbacks, but one look at it and you somehow get the idea you're not being taken. You simply don't buy such a vehicle to cruise down an Interstate at 70, or to com-

*This is a brutally tough machine developed by the British. Army in 1948. You can get parts for it anywhere in the world.*

pete with the racy lines of your next door neighbor's Aston Martin. It's intended for use in the back country and it makes no bones about it. For one thing it's the most widely known and universally respected four-wheel drive vehicle in the world, so it doesn't have to make any excuses.

We at *Popular Imported Cars* first became intrigued with the Rover back in the early 60's, when they began appearing in the United States with regularity. (About 1500 units are sold here annually.) In conjunction with a five-part test series on four-wheel drive vehicles currently running in our companion publication, *Auto Driver,* we finally got behind the wheel of the legendary vehicle.

Seating is surprisingly comfortable and the dashboard offers a full complement of easily readable dials and toggle switches. It is perhaps the best ventilated vehicle we've ever been in.

The body is of aluminum and heavily galvanized steel, so the unit is virtually rustproof. The little four cylinder engine, which delivers 124 foot-pounds of torque to an absolutely fantastic set of gears, is made of forged steel, cast iron and aluminum alloys. The gas tank has a 12 gallon capacity, but fuel consumption is good for a 4wd—about 15 mpg—and an auxillary 7 gallon fuel tank is offered.

Somewhat hampered in the back country by a 38 foot turning radius, the Rover makes up for it with a very tight steering box, superb suspension and its gearing set-up. The main gear box is a four-speed (first and second aren't synchroed) which is coupled to a two speed transfer case. Once you **CONTINUED ON PAGE 95**

*The brakes took this water bath and stayed remarkably dry.*

# Adventures in a New

# Guinea Rent-a-Wreck

*Peter Fitzpatrick goes bush in Nuigini and discovers a new missionary position...*

RAIN, RAIN, RAIN! Boredom, boredom, boredom!

A long weekend stuck in a motel room in Port Moresby. The casual suggestion it would be nice to go surfing. Vague memories of a previous trip down the coast, a village called Hula, night fires on the beach, guitars, baked tuna, surf, and we're off, turning the town upside down for a four-wheel-drive to hire.

Finally a khaki Landrover of dubious vintage is located at the Dive Shop in Badili. Air mattresses are piled aboard with a few tins of bully beef and we head off through the Moresby murk squashing cane toads and leaking profusely.

The first part of the drive is easy going. We pass villages with little fibro school houses and limp flags on white masts and painted stones at their bases. A big bridge over one of the estaurine creeks is crossed and we pass through dank, dark and dripping rubber plantations. A weak sun glows dimly in the drizzle and turns the long grass by the roadside a lamplight yellow. In patches of jungle snub-nosed and decaying World War II trucks nestle among the foliage.

After a few hours we reach Kwikila.

There is a new hotel but the town looks pretty much like it did three years ago. Council chambers beside the oval, a row of low cost housing to one side and the monolithic tradestore on the other side. A few people are sitting under the tradestore verandah chewing betel nut. We wave and disappear, turning left at the road junction.

We leave the gravel road and turn on to a dirt track. We drive for a while and then halt. We have no maps and there are tracks everywhere! Memories are wracked and we proceed cautiously. An old man appears on the track and Bob leans out of the window to ask directions. The man scratches his head.

"Lau diba lasi," he says, "lau diba lasi."

Geoff crawls out from amongst the luggage and asks directions in Motu. The man feigns ignorance. Geoff sighs.

"He wants a lift."

We shrug and the old man climbs aboard, betel-nutted teeth flashing amid vague Motuan promises of acquiescent granddaughters and surefire surf and sunshine.

The track gets windier and wetter. We enter an estaurine area and the track disappears into a sheetwash of rain and saltwater. We proceed gingerly and eventually climb out of it and on to a plateau of orange mud. We discover a new phenomena — no traction! The Landrover, in low range and second gear, skates on a greasy orange film as its fancy dictates. Gordon eventually masters it and we sashay off in the direction of Hula.

We drive through a sorghum field and enter dense jungle. More sheetwater appears and then the track rises out of it again. We eventually reach Hula. There is no sunshine and the girls are out to lunch. In fact everyone is out to lunch. A desolute pig sloshes across a clearing between tin and fibro houses.

"Weren't the houses made of limbom with sac-sac roofs last time?" puzzles Gordon.

"They were three years ago," says Bob, "I guess this is progress!"

The old man quietly climbs out of the vehicle, nods this thanks and disappears into the drizzle. We drive around. Finally we pull up near the beach. I climb out of the cab. A striation of driftwood, coconut husk and tin cans slope off the windswept beach and into the sea. A covey of rusty galvanised iron houses on stilts cluster in the distance.

We drive through the village and stop at the tradestore. A rumpled European in a dirty lap-lap emerges from another room and a din of radio music. We buy cigarettes and the trader asks if we are returning to Port Moresby. He suggests we seek accommodation at the school for the night and drags a nylon mail sack from under the counter.

"The mission's mail," he explains, "we haven't been able to send it out for a week now."

We climb into the Landrover, tossing the mailbag into the back and head for the school. The single teacher is happy to see us. He provides lunch and opens up a big fibro house on the edge of the school for us. At the end of the oval three fibro classrooms with low walls and no windows stand on the edge of the beach. A row of tall coconut palms are lashed in the wind and rain. We wander over to the classrooms. Holes punched by stones and feet litter the walls. Beyond there is surf — wild, ripping and rain lashed. We return to the house and soon the teacher appears with an offer of dinner.

Over the meal he explains the change; the young people leaving for jobs in Port Moresby; the ones who remain resorting to frustrated vandalism; the ones who find work salting their wages into fibro and galvanised iron. On a happy note he describes the sun and surf that still pounds on to the coast in the dry season. We sit up late drinking his claret and finally retire to damp mattresses in the big house. In the morning we set off for Port Moresby.

We drive through the jungle and splash through the returning tide before reaching the sorghum field. The Landrover slides off the road occasionally and we manhandle it out. The field gets muddier and the water deeper. Finally the Landrover bogs. We climb out. The chassis and sump are firmly wedged on the hump in the middle of the track. The track is flowing like a miniature creek and cutting the clay away from under the wheels. We rock the vehicle while Gordon plays with the gears. Finally we jack one side up and heave. The Landrover slides off the jack and rocks gently on the hump. Bob sits dejectedly in the flowing track. Geoff begins to harvest stalks of sorghum to stuff under the wheels. We free the Landrover and proceed at a snail's pace over the newly harvested bed of sorghum.

"Probably belongs to some rich planter," mumbles Bob in mitigation.

Reaching the end of the field we proceed cautiously. The black soil begins to give way to yellow clay. We bog the Landrover again but this time in sticky orange goo. We dig-heave dig-heave for an hour with no result. We sit smoking under a tree wondering what to do. Someone remembers seeing a small village somewhere nearby. We walk up the track. After half an hour we see the telltale coconut palms off to one side of the road. We wander into the village.

"At least this one looks like a village," Bob comments as we take in the elongated beehive houses of limbom and sac-sac and the carved platform in the corridor between them. Geoff approaches the first house and knocks politely on one of the stilts. A face appears from the smoky interior. Geoff explains our predicament. The man makes apologetic gestures and indicates the black clouds gathering for another downfall. We stand around and he finally climbs out of the doorway. He disappears into another house and reappears shrugging his shoulders in the same gesture. A bevy of children gathers under the nearest house to stare at us. We offer the man money. Still he shrugs. We give up and return back along the track. After a hundred metres a cacophony of children skitter through the mud to catch up.

We rock the Landrover, push the Landrover and berate the Landrover until it finally relents and pops out of its muddy bog. In triumph we splatter back into the village, a veritable travelling mass of mud stained children. Geoff and Bob pound on the shuttered counter of the village tradestore until a

# ADVENTURES IN A NEW GUINEA RENT-A-WRECK

disgruntled man with many jingling keys appears to open it up. We buy his complete stock of sugar, biscuits and sticky tin milk and hand it over to the kids. One of them races off for a can opener and the others retire to the jungle of house-stilts to divide the spoils. We climb into the Landrover and splatter on our way.

As we slide round a bend a little group of villagers with suitcases and parcels trudging heads down into the rain come into view. We stop, make enquiries about their destination and apologise for not having room. A grizzled old woman appears at the cab door clutching a younger woman by the arm. She gestures and gabbers until a young boy comes to her aid with a translation. The young woman is a nurse. She has to be in Port Moresby the following day. We confer briefly and I open the passenger door.

She climbs into the cab and over the seats, leaving dainty orange toe prints on Bob's shoulders. I glance up as she disappears into the maw of luggage and notice Geoff's eye shining in the gloom. There is a brief shuffling noise, a giggle and then the monotonous creak of the Landrover sloshing through the mud.

We swing around another corner and there is a creek. The track disappears into an ominous bulb of turgid yellow water. Somewhere below the muddy water there is a ford. We know because we crossed it yesterday. We dismount. There is no sign of Geoff. I gingerly wade into the water. After fifteen minutes I think I know where the ford is hidden. Gordon edges the Landrover into the creek. Bob and I, prompted by a mutual fear of vehicles in flooding creeks, stand up to our knees in the water.

The vehicle slides slightly sideways into the water. Gordon, in second gear, guns the engine. The Landrover lurches forward and stops suddenly with its nose in the air. The engine roars and a fountain of yellow water climbs up each side. Bob and I throw our weight behind the vehicle. Water shoots up my shorts and nose. The Landrover screams and levels out. We heave against the inertia until it finally moves. Gordon drives up the other side of the creek and squelches to a halt. We remount.

The engine roars. I climb out again and look under the vehicle. No muffler or tail pipe! I wade into the creek again and probe with a naked foot. Miraculously the tail pipe and muffler are wedged against a submerged rock. I retrieve it and trudge back to the vehicle. I consider the implications of reattaching the muffler and finally toss it into the back of the vehicle. A soft grunt is the only reply.

Finally we reach the gravel road and chug and gurgle into Kwikila. A petrol refill and we are on our way to Port Moresby. The first stop is the hospital. The Landrover rattles to a halt in the car park shedding layers of orange mud everywhere. A shy smile and the nurse climbs over the seats and waves goodbye. Geoff's head finally emerges from the depths of the luggage and he waves in return. We head for the Dive Shop sounding vaguely like a low flying jet.

Once over the hill and on the final run into Badili Gordon cuts the engine. We drift silently into the driveway of the Dive Shop. The proprietor emerges rubbing a greasy towel between his hands.

"Lost the muffler chaps?" he says.

We return to the motel. The following day we catch a taxi into Boroko. On the outskirts of Badili Bob nudges me and points to the roadside. A thin wisp of air whistles up my nose. A twisted mass of khaki lies scrunched under a tree. The name of the owner of the Dive Shop is clearly visible on an upturned and crumpled door. We stare in silence.

# SECOND HAND BUYERS' GUIDE

# LAND-ROVER 2.6

Readers often ask why we don't do road tests on vehicles somewhat advanced in years, as a service to second hand buyers. But there are problems herein, as each driver's driving style and demands will affect the shape of the vehicle and its components after thousands of kilometres. However, LES JAMES 'attempts the impossible', and brings us a roadtest on his tried and trusted 2.6L Land-Rover.

THE LAND-ROVER six cylinder 2.6 litre engine is now out of the civilian line-up of LWB Land-Rover vehicles. Replaced (wisely?) by the 3.5 litre V8, its death was barely noticed, except perhaps by followers of the marque. Some didn't even know it had ever existed while many six cylinder owners wished it never had. It was a motor that never really took off with Landy owners, for many felt there was little point in buying a motor that was only slightly bigger in cubic capacity, weighed more, and wasn't as strong internally as the four cylinder motor.

Competition from the then rising sun Toyota Land Cruiser forced Leyland to select the 2.6 litre engine for inclusion in the line. The motoring population was looking for more power and a better cruising speed. Whilst the 2.6 has good low down torque, a long stroke motor is defeating when looking for better highway speeds. At the time it was the only motor in the Rover range that was compatible with existing driveline components.

The intention of this story is not to provide a Leyland propaganda sheet but to state the facts as I have found them in the life of my own vehicle a Series 2A (1969) six cylinder LWB Station Wagon. Whilst I confess to an emotional attachment to "Old Red" I have also had plenty of opportunity to witness its strengths and weaknesses. Now in its 13th year, I can confidently say it has suffered few of the problems they are alleged to succumb to. (Perhaps it was all a Japanese plot).

I started travelling in a VW Country Buggy. It was a splendid little vehicle but not ideally suited to long distance outback overlanding. From the VW to a Falcon ute; the ute proved its worth but with more emphasis being placed on 4 x 4 travel, had to go. The deciding occasion was when it fell on its side whilst trying to negotiate a four-wheel drive track. But what to buy? I liked both the Patrol and Land Cruiser, however both vehicles at the time had threespeed gearboxes. To me this was a very negative point. I had been driving Land-Rovers in the Army while doing Nasho and seeing the use and abuse they withstood, I settled on the Rover.

From then until time of writing the red Rover has travelled over 198,000 miles. (It has an imperial speedo and odometer). At 162,000 miles it became a matter of 'top up the oil and check the petrol.' So I purchased another 2.6 second hand motor for $150.00 and rebuilt it. The reason for buying another motor was for future spare parts. For example, an alloy head costs around $600.00, so $150.00 for a complete motor became a good investment. The replacement motor components such as crankshaft, con rods, piston skirts and flywheel were all balanced by Paul England of Pascoe Vale, Victoria. With the cost of balancing, honing of the block, crack testing, new parts, (i.e. rings, bearings, gasket set), $1,200.00 was the final figure from purchase of the second hand motor to its installation.

The balancing has certainly been appreciated, as its quietness and smoothness have paid dividends in performance. Working on the 2.6L when on a workshop bench is a very easy task, but not so when the engine is *in situ*. The exhaust tappets can involve many skinned knuckles and oaths.

The old motor was stripped for examination and it was found that the cylinders needed honing, (not reboring); the crankshaft needed a regrind; the valves were not burnt but obviously needed a regrind. The valves had been reground twice previously in its 162,000 miles. Not once has the old motor or the new one burnt a valve. So why have other six's achieved the valve burning name? I believe there are two main reasons.

1. Early six cylinder motors were supposed to have their tappets done to 12 thou. clearance. This is far too tight. 15 thou. is more like it.
2. Many people run their vehicles on super grade fuel. That's fine if the engine is tuned accordingly. I run mine on super grade with a fuel air mixture set *slightly* lean, thus reducing the possibility of burnt valves.

Other than the motor the only other major expense has been the gearbox and transfer case. Both these were overhauled when the engine was done. Unfortunately I couldn't afford a complete job. So three years later it had to be completed. It is always unwise to use new parts with old. It would have been cheaper in the long run to completely rebuild the gearbox and transfer case all at once. The end float in the shafts, first, second and third gears and bearings all needed attention. Whether it is generally considered reasonable for a 10 year old gearbox and transfer case to require overhaul is probably debatable.

Considering the hard work it had done I personally felt it to be reasonable. Possibly the greatest surprise to readers will be the fact that the original Rover axles and diffs are still in the vehicle. I have carried 2 spare rear axles for many years but never used them. The rear diff at 198,000 miles is now showing signs of excessive end float although the bearings are not noisy. The front diff has had free wheeling hubs most of its life so it is still in good condition. Whether the rear diff will be replaced with a Salisbury unit or another Rover diff has not been decided.

The only problem ever encountered with the drive line was a left hand drive member. The drive member's splices disintegrated at slow speed on a bitumen road. To this day the cause has never been detected. The diff pinion seals have been replaced twice in the rear diff and once in the front. Whilst on the driveline, one of the annoying things on the Rover axles is the rubber backed felt seals that fit over the outer ends of the axles. Sometimes they last for ages while others only last for a few months. Wheel bearings have been replaced once. Brakes have had 2 complete overhauls. All universal joints have been replaced once. The brake master, clutch master and clutch slave cylinder have been replaced once. With the exception of the drive member all parts have been replaced due to normal wear process.

Another bone of contention for early six cylinder owners has been the electric fuel pump. On the 2A model it was a double entry electric pump. In severe heat conditions it can break down. However, breakdowns usually occur when the pump needs servicing. Most people are prepared to put new points in their distributor to keep the engine in tune but they seem to neglect the fuel pump. This model has a set of points each end. They can be serviced by the owner with care and patience. It is preferable to have it done professionally as the magnetic lift can be checked at the same time. My pump is usually serviced every two years by Wilsons Carburettor Service of Carlton, Vic. at an average cost of $35.00. The 175 CDs Stromberg carby has never failed. It is serviced regularly and to date the only major parts required for replacement were the throttle shaft and butterfly. With the addition of long range fuel tanks, the fuel system has an inline filter for each of the 3 tanks, a water trap filter before the pump and another inline filter before the carby. Despite the severe dust conditions "Old Red" has been in, the standard oil bath air cleaner has never let the engine's breathing down. When in severe dust conditions the filter is cleaned daily and the oil changed daily or weekly. Otherwise, normal servicing is followed.

Aside from normal maintenance and usage costs (tyres, petrol, oil, registration), around $5,000.00 has been spent on this vehicle, on the engine and transmission rebuild, refurbishing interior and exterior, and additional equipment/accessories.

For its working life, dependency and contribution to enjoyable leisure time, $386.00 per year is not an unreasonable figure. The extra equipment/accessories include for the interior: tachometer, combination water and oil temp gauges;

as well as battery condition and amp meter. Then there's a first aid kit and two fire extinguishers (1 front, 1 rear). The right hand rear wheel arch has a built-in cupboard and a net is suspended across the rear of the roof area for storage of maps, (also useful for drying damp clothing while travelling). It was a ten seater originally but has been converted to 4 seats. The rear bench seat was removed and a steel frame made to accommodate two bucket seats (ex Gemini). This frame sits on existing hinges and mounts allowing the seats to be folded forward for additional load or sleeping. Between the two rear seats is the Koolatron fridge. The fridge sits in an alloy cradle and is secured by a lap seat belt. The front bucket seats are ex-Fiat prime mover. All seats are reclineable and the mounting was approved by the police vehicle examiner. The roof and doors are lined with 25mm polystyrene foam for acoustic and thermal insulation. An overhead console with Pioneer cassette completes the interior.

## Exterior

The rear wagon windows were the sliding type as are the front ones. However, they had a continual habit of rattling loose and sucking copious quantities of dust. They were removed and replaced with tinted, laminated and fixed one-piece windows. The body from the roof line down was sprayed in Dulux enamel with an added polyurethane hardener. After weeks of sanding and preparation the five top coats were sprayed on. The side door sills were removed and a side scrub bar replaced them. The original front mudguards were replaced with army style cut out guards. Most of the vehicle's early life was spent in the high plains in Victoria in snow and mud conditions. In these conditions the army guards give excellent self cleaning properties with no build up of mud. They also make life a little easier when working around the front of the vehicle. A tray was made up to accommodate the auxillary 90 amp deep cycle battery and recovery gear. This tray fits in between the front of the chassis in front of the radiator/headlight panel. An army style brushguard is mounted on the front bumper bar providing a mounting area for 2 Cibie 200 watt driving lights. Sealed beam headlights on relays replace the original lights, and there is a small vice and an antenna for the Royal Flying Doc Radio. The R.F.D.S. radio is from Allcom Pty. Ltd. W.A. Around the back two small bumperettes were made to replace the standard Rover ones. The Rover bumpers are really only suitable for lashing points if the vehicle is being shipped — as useful as an ashtray on a motor bike. Wheels are standard Land-Rover; tyres on the drive are Dunlop Triple Traction 8 ply and on the steering Bridgestone Jeep Service 6 ply. The reason for the mixture is the Jeep Service do not seem to have as strong a sidewall as the Triple Traction even though 8 ply Jeep service are available. The Rover tows an ex-army trailer on extended trips so the extra sidewall strength becomes important. Overall, the Bridgestones outlast and outperform the Dunlops.

With the money spent on "Old Red" a cynic might say the vehicle is over capitalised for its true worth. And possibly he would be right. After all a 13 year old 4x4, regardless of make isn't worth a lot of money. But judging by its past performance it should last another five to six years. Because it doesn't have the latest body stripes from Japan with matching whiplash suspension doesn't mean it can't keep up with the latest offerings in off-road performance.

As well as outlast them. Whatever happens to the Landy six now in the vehicle it will not be replaced with a Holden motor. The logic of using the Holden motor completely escapes me. The torque and power is wrong for the Landy gearing; the Holden distributor is set too low for sensible water crossing height, likewise the fuel pump. Many people who have converted to Holden power criticise the Rover's gearbox and diff for being weak. Perhap we could reverse the situation and place a Land-Rover motor in a Holden car and watch as the front end collapse. Would people then criticise Holden front ends for being weak? And while on the offence, has anybody ever compared the amount of broken Land Rover axles to the amount of cracked Land Cruiser transer cases? Criticism of other products in defence of the Land-Rover is a poor way of justifying the vehicle but it does serve a point so often overlooked. All vehicles have weaknesses. The key to their successful operation is being aware of them and not driving them beyond their ability or your own. So many Land-Rover axles have been broken by overloading, poor driving and wrongly matched motors, driving gearboxes diffs and axles that were designed for different vehicles.

Over the years I have come to know and respect the 2.6 and despite my admission of emotional attachment, I have based my admiration on the vehicle's performance. The only time in 13 years it has ever let me down was when the drive member failed. And while talking of failures I've developed an intense dislike of the firewall. It has cracked numerous times. Each time it cracks it is welded. I have been told with little comfort that six cylinder firewalls are prone to fatigue cracking.

On the credit side, the lack of power and cruising speed has never bothered me. Over a long highway run I am content to drive at a consistent 55mph (in overdrive) while the Landy averages 20mpg. The engine balancing and overdrive helps achieve this figure. The supposed lack of power has never stopped the vehicle from getting through any 4x4 conditions (from the snow and mud in the High Plains to the Great Sandy Desert). It would be different if you expect power to tow a large caravan or boat. Once again, it's a matter of the right vehicle for the job at hand.

Though I mourn the passing of the 2.6 litre, the new V8 and 3.9L Diesel will certainly give Landy buyers and knockers one of the most capable 4x4s ever. Leyland's hope of the new motor line-up giving the opposition some stick is probably well founded. It's a pity it's 10 years late. Goodbye 2.6.

# FROM POST TO PILLAR

We come across all sorts of vehicles in our travels, but rarely do we find one which turns so many heads... Overlander spoke to the owner of this Landy that had ventured far from its natural habitat in the Top End.

IF YOU'VE ever taken on the task of building or rebuilding a vehicle, you'll know it's a time consuming, expensive, and extremely frustrating business. If you haven't, as Neville suggests, then you're very wise.

And Nev should know ... he has recently rebuilt a very unusual Land-Rover. The vehicle, a 1972 Series 11A Forward Control, was in its day form a table-top truck ... Neville understands it was a PMG vehicle in its day ... but it has since undergone considerable facial and internal surgery. When 'Bigfoot' (named by Neville's children after its 12 ply 825 x 16 tyres) was found, it was virtually a wreck.

Two friends, however, came to the party ... Peter Lanski of Off Road Autos added his wealth of experience in off road vehicles and provided the mechanical expertise while Rick Maddox from Rick Maddox Engineering covered the structural and panelling aspects.

The vehicle consisted of the original truck cabin with the back removed. It had a station wagon body married to it from the passenger doors to the rear wheels. From there back a panel van body rear section has been attached ... sounds complicated!

The vehicle was completely stripped of everything; engine, transmission, running gear, floor and roof. The passenger doors and door struts were rusted out and were removed ... the doors renewed and new box sections made and welded to door struts where required. Special vents from a bus parts company also went into the front doors.

There were virtually no cross members supporting the floor so new square steel supports were welded in and a new 16 gauge steel plate floor laid. As well, wall panels were removed where straightened or renewed where required.

Before the flooring was completed a Holden 202 engine was installed and coupled with a beefed-up hydramatic/automatic transmission. The standard transfer case was separately attached to a cross-member and connected to the auto by a short drive shaft.

Rick Maddox then applied the cunning he has inherited from all his experience and designed this area so major overhauls can be done without demolishing the cabin. Structural steel around the cowling was bolted rather than welded ... so with the removal of half a dozen strategically placed bolts the flooring can be easily lifted away and the engine removed or replaced. As is the human way, "There was one thing I forgot to do," said Neville Mitchell. "The seal behind the crankshaft can't be replaced without removing the motor."

But on with the show ... the running gear was all replaced or overhauled, brakes renewed and a new booster unit installed, wiring renewed, and a paint job applied.

Inside he fully carpeted the vehicle with underfelt and around the cabin insulated with polystyrene, which makes for a surprisingly quiet ride in what originally must have been one hell of a noisy beast. Neville applied a greater level of instrumentation than is normally available in Landies, and installed a four-way speaker system. Up front there are Recaro seats, while in the back for the kids he's got European lay-back car seats. And there's still space to sleep his family of five.

With all this steel plate work and additional creature comforts you could be excused for thinking of Bigfoot as a heavyweight contender, when in fact it weighs in at 2500kg on the road and loaded. And despite, dare we say, the Landy's clumsy appearance, the vehicle is not top heavy. All the weight is down low, providing a safe centre of gravity to work with. The vehicle's height provides excellent ground clearance and escapes one problem faced by most other Land-Rover drivers ... dust. All because of its height.

On their first test of the vehicle, taken hardly before the paint had dried, the Mitchell family covered more than 9000 kms to Sydney towing a Chesney campervan. They clocked 22 litres per 100km (13 mpg) and encountered a few mechanical problems, but nothing a good man with the help of off road specialists in Melbourne and Sydney couldn't handle.

The vehicle will now be back in its home territory, around the Top End's rivers and lagoons ... on their way they stunned quite a few onlookers, which did wonders for Neville's pride. And all those hours stretching over 18 months of toil ... he puts that down to experience. Would he do it again? ... "Not bloody likely mate!"

*1963. Built for Shell Oil Co., for oil exploration on the Alaskan pipeline. The tyres and wheels cost more than the original price of the Land Rover.*

# Land-Rover—Transatlantic Specials and Prototypes

## by Michael Green

As the title suggests, this article covers a number of special and prototype Land Rovers which were built in the USA by the Rover Motor Company of North America during the sixties and early seventies.

Currently all Land Rover and Range Rovers are in great demand, and they have been modified, utilized and conscripted into various roles throughout the world. The following covers a number of special conversions, all of which were associated with the USA company, and designed and constructed by various local companies in conjunction with the RNA engineering staff.

In September of 1960 the late Donald Campbell arrived at Bonneville with the "Bluebird", powered by a Bristol-Siddeley "Proteus" jet engine to attack the world's land speed record. Included in the entourage were five special Land Rovers; an 88" converted by Minimax to a fire fighting appliance, capable of extinguishing jet fuel fires; two 109" truck cabs carried special equipment in the form of extra heavy duty 24-volt batteries, an 8-kilowatt generator driven from the centre power take-off, 100 psi compressors to feed portable high output cooling fans; and an instrument board mounted on the rear to control the various pieces of equipment.

Since the "Bluebird" had no on-board starting system, one of each of these tenders were stationed at each end of the course, start-up was by remote control via heavy duty cables inserted into a built-in plug in the body of the record breaker. The compressors supplied the air for the built-in jacking system, for at the conclusion of each run it was necessary to jack all wheels clear off the ground and rotate them to prevent the brake pads adhering to the discs. The surplus air was fed to two large fans (portable) which blew through ducting to cool the two transmissions.

Of the other two 109's, one carried electronic measuring instruments, by which it was possible to record speed and air temperature, both into and out of the jet engine. It was also possible while the "Bluebird" was in motion to review the operating temperatures of the transmissions, brakes, fluids, etc., whilst the remaining 109" was a special mobile workshop, constructed in Canada. All support vehicles with the exception of this, carried high frequency Pye radios and the two tender vehicles were also equipped with the latest type hydraulic drum and cable winch with a 4,000 pound line pull.

The Snow Rover was developed by J. D. Hopping of the Rover Company of North America. Built on an 88" truck cab chassis, a special adaptor was made to install a standard 2¼ litre engine in the load area to power the snow-blower, the normal hydraulic snow plough being driven from the front power take-off.

In conjunction with "La France", a special fire fighting vehicle was constructed, based on a 109" pick-up, for the J. C. Pratt Co. Both front and rear power take-offs were utilized to operate the pumps. Additional cooling was provided by an oil cooler working in conjunction with the fire pumps, enabling

*The Snow Rover developed by J. D. Hoping of the Rover Motor Company of North America. This had a 2¼ engine in the load area to run the snow blower.*

92

*The "La France 109" Fire fighter built for J. C. Pratt Co., with PTOs to run both pumps. These are still in use today in Hawaii.*

long periods of pumping to be maintained. During this period there were three other pieces of fire fighting equipment, built on 109" and 88" chassis in association with Western Fire Fighting Equipment, Inc. in San Francisco. The long wheel base is still in service at the Mauna Kea Beach Hotel, while the smaller units are in service on Maui.

In early spring 1963, at the request of the Shell Oil Company, a vehicle was developed capable of crossing frozen tundra waste for oil exploration in Alaska. This conversion was carried out by Hi-Flotation Incorporated, then located in Monterey. To overcome the steering drag of the Hi-Flotation wheels and tyres, a Garrison power steering unit was driven from the front power take-off. The centre power take-off point was used to drive a compressor to inflate the tyres, whose recommended pressures could vary from 2 to 12 psi. The cost of this conversion was actually more than the original retail price of the Land Rover!

The Rover Motor Company of North America (Canada) met the Canadian Forestry Services' request by developing a special, high ground clearance Land Rover, capable of covering ditches, drains, boulders, etc. Notice how this model is crab-tracked so the rear wheels do not follow the front wheel tracks in difficult terrain. (Editors Note: The wider front track also gives chassis clearance to achieve at least some steering lock with the very large tyres.)

The dual rear wheel project was an idea sponsored by Bill Reno, then a Land Rover dealer in Colorado and a great enthusiast. Not totally successful due to axle shaft overload, coupled with premature hub bearing/seal failure.

During 1966 the Land Rover V8 project came to life; the brainchild of William Martin-Hurst, then Managing Director of the UK Company and J. B. McWilliams, President of the Rover Company of North America. The object was to install the newly acquired 3.5 litre GMC V8 into an 88" chassis without altering the general configuration, in other words, the frontal appearance would remain the same. But certain other changes could be incorporated — brakes, axle ratio, instrumentation and interior trim, this project being under the control of Product Development Engineer, Richard F. Green, in the Western Zone Office.

The body work modifications to install the V8 engine were carried out by Moeller Brothers in San Leandro, California and apart from the installation of the V8, there were other modifications. A conversion to 10½" twin leading shoe brakes to match the new performance, final drive ratios changed from 4.07:1 to 3.54:1, electrical system converted to negative earth with an alternator in place of the original generator.

The stock grey drab interior was changed to black, with a new first for the 88" series, an adjustable driver's seat. The complete fascia was reworked to incorporate instruments from a Rover 3-litre sedan. Finally, the whole rig was finished in a pleasant shade of yellow "Golden Rod", hence the factory code name.

The initial performance was such that it out-performed the then current production car, a Rover 2000 TC.

During the time "Golden Rod" was under construction, the experimental department of Land Rover Solihull were assembling a similar vehicle with automatic transmission, for comparison purposes.

After a brief test period in California, "Golden Rod" was driven by the Product Development Engineer to New York, the trip being accomplished in four days; the first night stop being Salt Lake City, the second night being spent in North Platte, Nebraska, the third in Gary, Indiana, and into New York. From there to Southampton and delivery to Solihull by the American staff member, where it caused a considerable stir! For these vehicles were originally rolling roads for their successor the Range Rover.

"Golden Rod" was eventually shipped back to the USA, sold to an East Coast dealer and finished up in the hands of a student at Colorado University in Boulder, getting nearer home at every move. Should anyone know of its whereabouts, would they please contact the West Coast Land Rover Owners Group at 7440 Amarillo Road, Dublin, California 94566.

**Michael Green**

*This dual rear wheel 88" Pick-up was an idea sponsored by Bill Reno, a Land Rover dealer in Colorado. The axle and bearing load was too great.*

*Built for the Canadian Forestry Services for covering deep ditches, drains, mud, boulders, etc. The track on the front wheels was wider than that of the rear.*

# LAND ROVER LtWt GOES V8

**ADRIAN WOOD describes a successful transplant**

After owning 3 ex-military 2As, a swb for roadwork, trials and comp safari's in 1976-77 and 2 lwb hardtops used for trans-Sahara in 1973 and 1978 I decided the next Land Rover had to have more power.

I dismissed the Rover 2000/2200 engine as being a lot of work for very little power increase. The Ford 3.0 V6 was a possibility, but I could not find a conversion kit in March 1981. A kit was essential as I had normal tools, and an engine hoist, but the welding equipment was three miles away at work.

So it came to a Rover V8 3500 car engine as it was readily available at a reasonable price, and Phillips did a conversion kit for it. Alternative Rover V8's ie: Range Rover, 3.5 saloon/coupé or SD1 V8 are available, as is a choice of flywheel/clutch and exhaust, also the model of Land Rover to fit it in.

The following is an account of how I fitted a 3500 engine into a 1972 2A LtWt.

After removing the 2¼ litre engine and gearbox in the normal way, the gearbox was u/s, I had it rebuilt with a ground down first motion shaft. The V8 has a smaller spigot bush than the L/R.

The rebuilt gearbox was fitted onto its mountings and the conversion housing screwed onto the engine. A new bronze spigot bush was fitted in the V8 crankshaft. A Rover 3500S car flywheel bolted on with 3500S bolts and the clutch assembly fitted up. (Austin taxi pressure plate with redrilled dowel holes and std 2¼ L/R disc.)

The starter motor had to be filed up to clear inside the housing before it fitted flush. The front engine mountings were bolted on and car rubber mounts used. The engine was slung on a gantry and offered up to the gearbox, so I could mark the bulkhead to cut it to get the V8 in.

The kit instructions tell you how much to cut but as it varies for different models I decided to measure it myself. I cut the bulkhead with a drill and padsaw leaving 1" clearance between bulkhead and exhaust manifold and fitted the engine onto the bell housing. The front engine rubber mounts fit straight onto the existing mounts on the chassis.

The clutch slave cylinder bracket needed modification with a 2lb hammer to dish it sufficiently to fit around the housing. The clutch flexible pipe runs close to the o/side manifold so I slit 6" of heater hose to insulate it and stop the fluid boiling. The hose or sleeve must not touch the manifold.

I used a 15 ACR alternator mounted 5" higher than original, or alloy angle brackets and a QH 1175 vee belt. (I used a 15 ACR because I had one. A 16 or 18 ACR would be preferable for the electrical load it now has to carry). With the alternator mounted high it is possible to retain the std L/R battery position, but I chose to fit a thinner battery and mount the remote oil filter vertically on the side of the alloy angle brackets. (Remote oil filter is used as the filter on the V8 is just above the front diff.)

The standard L/R radiator was refitted as it had an angled bottom outlet, the straight outlet can be very tight to clear the oil pump and Hoses used were Renault 5, sleeved to fit the smaller L/R connections, with the top hose drilled to take a metal tubeless valve, bolted in minus valve core, to take the inlet manifold connection. All electrical connections were remade to suit the V8 and the std L/R choke cable was shortened to fit and a universal throttle cable used direct from the pedal to carb linkage, to retain normal pedal travel. The V8 mechanical petrol pump was used, with the carb return to tank blanked off. The 3500 air cleaner would not fit so it was cut up to make two separate air cleaners. Alternative air cleaners 3.5, RR or S.D1 with short elbows will fit some L/R's, but they all foul on my bulkhead, and there is not room to fit them at the front of the engine.

The exhaust system was made up by using as many std items as possible. On the n/side a cut, angled and welded manifold, short RR downpipe sleeved onto a shortened military front pipe, with std intermediate pipe and silencer on std L/R mountings.

The o/side was more difficult but using a shortened bent and cut military front pipe it passes through the clutch crossmember, running in front of gearbox and bent to run parallel to the n/side pipe and in to a shortened intermediate pipe with a L/R silencer mounted on diff side of existing box — both tailpipes are bent to miss o/s/r tyre, but they are more or less in the std position. The pipes are not interconnected in any way, none of this exhaust is lower than the original, which is essential if it is not to be damaged off road.

A new transmission tunnel was made from sheet metal fitted with self tapping screws to the bulkhead and floor.

The engine-driven fan runs too close to the radiator for reliability, so a pair of electric fans were mounted in the radiator recess. The engine does not run hot on the road. But the fans are needed for town or off road. As the engine is mounted high at the front the front prop runs at a steeper angle than std, so the u/j was split and the shoulders and flange cleaned up on the grinder, to give slightly greater clearance at extreme axle travel.

Michelin 750 x 16 XS mounted on lwb rims and 3.54 ratio diffs fitted to raise the gearing with the speedo recalibrated. Corbeau seats, auxiliary driving lights and air horns were considered essential items.

What about the brakes you say! LWB 11" front brakes with servo really are essential as the std 10" are not up to any increase in performance.

That is the way I did it 2½ years ago. The LtWt has been reliable for 16,000 miles including recovery marshalling at 19 A.W.D.C. comp safaris, towing trailers and general Land Rover type chores, it has been reliably checked at 100 mph but cruises mainly at 70 mph + and with normal usage fuel consumption is better than std for a faster road speed.

The disadvantages — the transmission was not designed for 140 bhp and must be treated with respect. The heat build up in the cab in summer traffic even with soundproofing/insulation on the floor and bulkhead is almost unbearable, I keep thinking about fitting footwell vents.

If any one is thinking of doing a similar conversion and needs help/advice give me a ring on 086735-577 evenings or take a trip to Wycombe Pub Meet — it needs your support.

---

## LAND ROVER

**CONTINUED FROM PAGE 83**

get the hang of double-clutching, you can drag two tons of dead weight up a hill and out onto level ground just as though you had an eight-speed main box.

The vehicle cruises comfortably on the highway at 60 mph, and fourth gear will let you loaf along over twisting dirt roads at 20 mph without straining or lugging. A 4.7 rear end and gearing for high torque/low speed situations (see box) limits top speed to 65.

Lack of speed and quick acceleration and a complicated series of laws which limits importation privileges to the short wheelbase, gasoline-powered model pictured here, have limited Rover sales. It is, at the present moment, impossible to buy the larger (109 inch wheelbase) station wagon model or the diesel engine models. Rover's 6 cylinder didn't meet government anti-pollution standards, but Rover has developed a new V-8 which we should see here in the states in mid-1970, probably mounted in the 9-passenger station wagon, a vehicle which should really set the four-wheel drive market on its ear.

For the present, the man who enjoys spending his spare time out where there are no roads, the little Land-Rover is a dream come true. One thing for sure — he won't have the family on his neck. It's a precision engineered getaway car with an all-male personality, built for the man who *really* wants to get away.

### LAND ROVER 88 HARDTOP
#### SPECIFICATIONS

| | |
|---|---|
| Horsepower: | 77 @ 4250 RPM |
| Torques: | 124 ft. lbs. @ 2500 RPM |
| Displacement: | 139.5 cubic inches |
| Compression Ratio: | 7:1 |
| Electrical System: | 12 volt |
| Tires: | 7.10 x 15 |
| Wheelbase (inches): | 88 |
| Track, Front and Rear (in.): | 51½ |
| Length (inches): | 142½ |
| Width (inches): | 64 |
| Height (inches): | 77½ |
| Turning Circle (feet): | 38 |
| Ground Clearance (inches): | 8 |

### MAIN GEARBOX RATIOS

| | Transfer Box | |
|---|---|---|
| | High Ratio | Low Ratio |
| First gear | 18.264 | 43.941 |
| Second gear | 12.438 | 13.398 |
| Third gear | 8.414 | 21.164 |
| Fourth gear | 6.110 | 15.360 |
| Reverse gear | 15.560 | 39.147 |

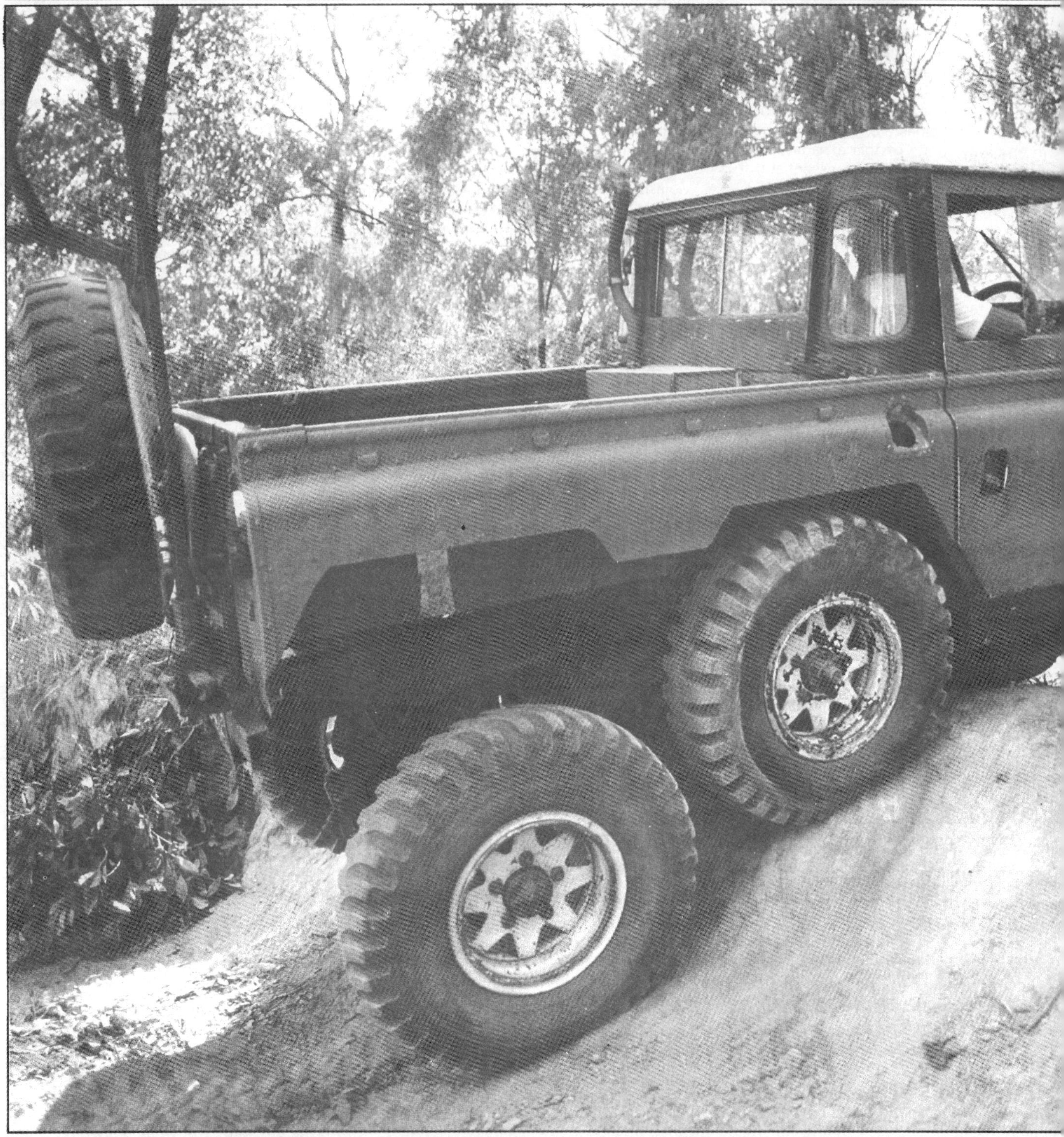

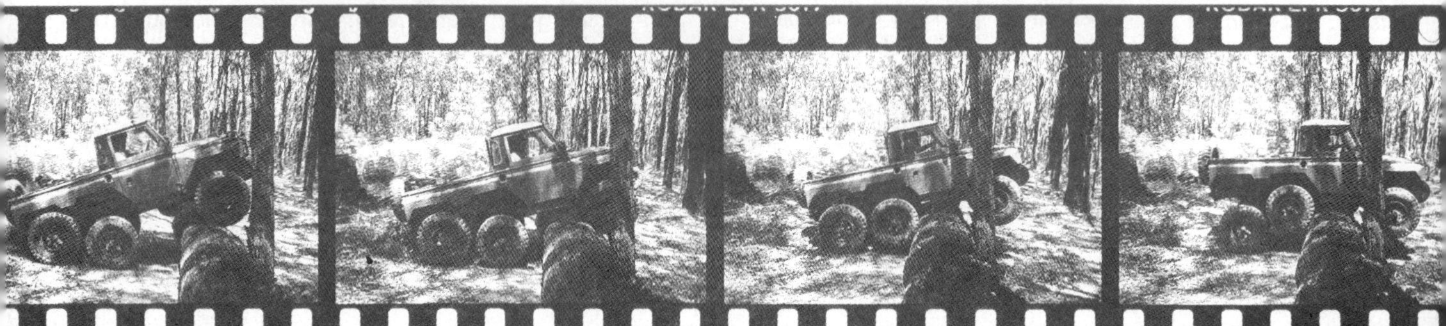

# LARMAN'S LANDIE!

An ability to carry large loads into out of the way places or to provide armchair-like comfort and limousine-smooth looks in town were probably the last things in this man's mind when he set to work on his idea of the perfect off-road machine. But all-terrain capability was another matter. Bob Maron takes a look at Bill Larman's back-yard 6x6 Land Rover.

Bill Larman spends most of his time working on other people's Land Rovers — reconditioning engines, engineering modifications and building virtually anything his customers want. He does all of this from his own backyard workshop, which is handy because, when he's finished the work that earns him a crust in the Melbourne outer suburb of Kilsyth, Bill can work on into the night on his favorite toy: a 6x6 Landie that is stunning not only in its effectiveness but also in the ingenuity it displays.

This is not Jaguar-Rover's 6x6, nor anything like it. Perhaps the closest thing to it available from a manufacturer would be a Pinzgauer 6x6, but even then there are extreme differences in intended role. Where vehicles such as the Jaguar-Rover product were designed primarily as high capacity load carriers for normal 4WD terrain, their long wheelbases, relatively poor approach and departure angles and lack of diff locks severely limit their effectiveness in true all-terrain conditions.

The Larman Landie is something else, designed solely to maximise its performance in difficult situations with very little consideration given to such things as highway performance, road handling, comfort or appearance. This thing will go anywhere — almost — and do it with utterly convincing ease.

It took Bill a mere six weeks to take his concept from the germ of an idea and build it into hard metal and alloy. But despite the already superior abilities of his beast, he has continued to refine and improve it; adding new ideas, altering others, continually in search of ways to make his 6x6 even more unstoppable.

The basis of Bill's handiwork is a Land Rover Series II chassis from a short wheelbase model, but all similarity with the vehicle's origins ends there, and with the bubble-top cab whose purpose, Bill claims, is merely to separate the wheels and mud from the driver.

Starting at the front, Bill's machine immediately gives hint of the incredible capabilities he has engineered into such a basic vehicle. Axles are shortened long Series I units, cut and splined to suit Bill's own carrier. Series I Tracta joints have been used as well, largely because Bill found that CVs or universal joints would rarely last more than a week under the immense loadings.

The entire assembly is suspended from the chassis by a centrally pivoted A-frame and Watts Link, bushed with a Kenworth torsion bar rubber. And that's not the end of it, because the axles themselves are hung from Firestone No. 62 airbags which can be pressurised from within the cabin to tilt the body up to 15 degrees from the angle of the axles. Bilstein shocks supply the appropriate damping.

Front differential is the same as the two at the rear: a 4.7:1 Land Rover, but at the rear it is the incredible extent of the wheels' articulation which is most astounding.

Drive for the twin rearward facing bogey diffs is supplied via modified Land Rover transfer gears which drop power to axle level. On the other side of these cases, two slip joints and universals have been employed on a short tailshaft to enable the shaft to operate at a phenomenal 45 degree angle to the diff housing!

The Salisbury rear axle half-shafts themselves use another of Bill's own design carriers and are sprung from the bogey with a single set of inverted Land Rover springs — the axles being mounted to the ends with the centre fixed to the bogey. Dampers keep things from bouncing around too much, but that isn't the whole of the rear suspension because the bogey itself is suspended from the chassis with torque rods, the result being some one metre of travel at each wheel!

From there, things return a little more to normal, though some of the figures are still staggering. The engine is, at the moment, only an old Holden 186, though plans are underway to fit a bigger Holden/ Ford hybrid — and that's a story in itself.

Bolted to the rear of the 186 is a four- speed International truck gearbox which feeds into a standard Land Rover transfer case with early model low range gears. Incredibly, and proving the strength of the British product, none of the transfer gears has failed yet even though they have to cope with around three times their design load. But cop these figures: low range first gives a ratio of 86.5:1 and low range reverse 105:1. Not surprisingly, Bill reckons he rarely has to go that low, preferring to use low range second for even obstacles that would be literally impassable in a lesser machine. Incidentally, Bill was responsible for the design of the first Roberts diff lock, and has one fitted to each axle — all of them individually operable from inside the cabin.

This is all pretty spectacular stuff (especially when you remember that Bill built it all with his own hands), but there is even more to it. Not satisfied with what was an 82 degree approach angle (with Ramsay 8000 lb winch fitted), Bill mounted a pair of 4.00 x 8 8-ply wheels to the front, one each side of the front-mount winch. Although that reduced his *measured* approach angle to 65 degrees, it effectively improves it, allowing him to roll up and over obstacles which present a full 90 degree face!

Of course, getting the front wheels up is one thing, but how do you get the rear end to follow? Well, for a start, the use of a bogey rear end means the first of the rear wheels is moved forward, creating an effectively shorter wheelbase and thereby improving the ramp-over angle to the point where, at 109 degrees, mere mortals could not conceive of the need for more. Wheelbase at this point is just 74 inches, with another 37 inches including the rear axle. However Bill found that the belly sometimes bottomed out when driving over logs — logs more than a metre in diameter. No, that's not a misprint: more than a metre.

You and I would perhaps consider such a limitation as meaningless — after all, how often are *you* faced with such an obstacle? Not good enough for Bill, I'm afraid — he decided there had to be a way of solving the problem and not only did he do so, he did it in such a way as to make the measurement of ramp-over virtually meaningless. What he did was to fit another set of those small wheels, this time to the belly, half way between the front and second axles. Then he fed them power from the engine through a gear train from the PTO that gives them the same circumferential speed as his main wheels.

Now, even though his measured ramp-over is reduced to 129 degrees, he can still drive forward even when all six main tyres are suspended in the air — like when he's balanced on one of those giant logs.

With all this mechanical ingenuity hanging from the chassis, you might be excused for thinking the beast's own weight would defeat it, but that's simply not true. The bubble top body is a mere shell (narrowed at the rear to cope with the phenomenal suspension travel) and all the real weighty gear is slung low — most of it below the top of the tyres. In other words, the centre of gravity is low; so low, in fact, that the vehicle has a roll-over angle better than 60 degrees! Even when he's using this ability to drive across a slope, Bill can do so in comparative comfort. With the downhill airbag inflated he can level out the ride to a more acceptable approximation of horizontal.

Those airbags assist in other ways too. Unpressurised, there's over 280 mm of air between the ground and the diffs, but pump them both to maximum and Bill gets 610 mm of clearance under the chassis rails.

In any case, the weight isn't that much of a problem. Remember it has six wheels, spreading the load over a 33 per cent larger contact area and, with just 6 psi inside the rubber during off-road work, the 900/16 bar treads can walk over soft sand or mud with surprisingly little sinking.

For our benefit, Bill consented to a demonstration of his Landie's ability. This was the second time I'd seen it at work, and I still can't get over the impression that the thing is unnatural. The photos hardly do it justice, but the thing simply nosed up to a four-foot vertical climb and went over. The nose wheels pressed against the dirt and squirmed upward, dragging the front drive wheels behind. After that it was just a matter of the rear end stepping up one axle at a time.

After that he attacked a gully that required a climb out that must have been all of 45 degrees and more, but there was barely any wheelspin as the machine heaved itself up the rise. Finding that none of this was proving to be any kind of a challenge, Bill drove along the side of the same gully, dropping into a deep bog-hole at the end. Now that was spectacular. He did it twice, the rest of us holding our breath, before a slight misjudgement caused the downhill front wheel to dig into the mud. That was a shame, because it caused the almost unthinkable to happen. It rolled onto its side.

Undeterred, Bill set about extracting his vehicle — a process which proved that the more capable your off-roader, the bigger the trouble you can get yourself into! Nevertheless, the 6x6 was soon upright again and Bill performed the whole thing again, explaining that, if you fall off a horse you've got to get back on quick, or you never will again.

The whole incident, far from detracting from the impression of unstoppability, only showed that Bill is no boulevard four wheel driver. He *uses* every scrap of his vehicle's capability. His 6x6 may have fallen over, but nothing else could even have got there.

It was after all this, with the vehicle still suffering damage from its unplanned "rest", that he demonstrated its amazing log-climbing. Who needs a chainsaw? Not this guy.

But, as I said at the start, Bill has hardly finished perfecting his vehicle. Already he is working on a hydraulic ram attachment that will enable him to lift either of the rear axles from the ground. The benefits are obvious — not only will he be able to choose his wheelbase, but lifting the rear axle will take a lot of weight off the front wheels, making it even easier to climb over even larger logs. Being able to convert to 4WD from six at the press of a button also means he can dial in the amount of weight on each set of tyres, giving him either a soft 6x6 footprint for deep sand or a heavier bite through slick mud on hard-packed surfaces.

God alone knows how he could possibly improve his Landie's off-road performance after he's fitted the ram, but having seen Bill at work, and knowing how unstoppable the thing is already, you can bet he'll think of something!

**4X4**

# LAND ROVERS

**CONTINUED FROM PAGE 76**

The Bomb could land right on top of it, but somehow you're sure that it would only blister the paint a little. A Land Rover is the wheeled embodiment of the spirit of one of the sturdiest, most indomitable nations on earth. This is not just a heavy-duty vehicle; this is *John Bull's* heavy-duty vehicle. And there's a difference. There it is—Rule Britannia and Press On Regardless!

Okay, yes, I'm sure your 4wd vehicle is just as good, if not better. But it can't have anywhere near the Land Rover's class. And when it comes to tradition, well . . . hang around Land Rovers long enough and you'll wind up convinced that if Rover ever stopped making them, the whole continent of Africa would sink like Atlantis into the sea. Anyway, two days after I started, I was a confirmed Land Rover nut. Those things are more fun to drive than anything this side of a Ferrari! They'll go over, under or through anything, the visibility's marvelous, and you have to be really creative to make one break. And for sheer startle value, a Land Rover just can't be beat.

When you took a customer out for a demonstration ride, you'd get him (or often, her) firmly strapped in and take off down the street, which was separated from an expressway by a rather steep grassy embankment about 10 feet wide, and during the winter, this strip always had snow piled up on it about three feet deep. You'd get up to about 20, say very casually, "Hey, why don't we take the expressway—it's quicker," and suddenly swerve right. As you swerved, you banged it into 4-wheel drive and "whumph!" Into the snow, churn up the embankment, pull out into the disbelieving traffic and go, hood and fenders festooned with hunks of snow, customer still softly going "gah . . . gah . . . gah . . ." to himself. Then down to the river where you'd demonstrate the thing's ability to climb sheer cliffs, charge through the woods, wade through hub-deep sand and generally do unreasonable things without a whimper. Then you'd let the customer play for awhile, and it was *your* turn to hang grimly on while *he* tried to destroy the car, the object being to let him get so intrigued that he could be relieved of his sack of coin. And oddly, there was only one instance of customer-caused damage to our demonstrator. And not in the bush, either. This happened in downtown traffic. You remember Land Rover's claim that the thing's built to withstand the full charge of a bull rhinoceros? Yeah, well, they'll even do better than that.

On this particular day, my prospect was one of those marvelous old ladies New England's so full of. There's a whole class of them; great, huge, jolly people with master's degrees and sometimes doctorates—always from Smith—and they're really into living. They do things like sail star boats single-handed, dig bushels of clams for dinner, march in demonstrations, lecture at the library and talk a fascinating blue streak while they get genteelly swacked on sherry. This one was the archetype of the species, and she was having a ball with the Land Rover, giggling and cooing as she howled through a trafficky, one-way circle in a beautifully-controlled drift. Then, ohmigod, here came a great Mother Buick the wrong way, and with a cataclysmic bang the two cars married fairly, front to front.

We fared a lot better than the guy in the Buick simply because we were harnessed firmly to the seats and he wasn't. He banged his head smartly on the windshield, sustaining Slight Injury, which rated us all an ambulance, a fire engine, all the policemen in the world and enough spectators to stop traffic completely.

Then there followed the required Great Flap about Who Was At Fault, wherein Brunhilde stood like a pillar of New Hampshire granite and told all and sundry concisely, precisely and politely that she was in the right and they could all go to hell, and finally, like the mules that drag deceased bulls from the *corrida,* the wreckers came.

Just for the hell of it, I got into the Land Rover and started it. It idled quietly, with none of sounds like the fan makes when it's stuck into the radiator, so I put it into reverse and tried backing out from within the Buick. It went backwards, alright, but the Buick wanted to come too, so I got a wrecker to sort of stand on the other car's tail and tried again. This time the Buick fell off onto the road, and I got out to inspect the damage. The Buick seemed utterly destroyed, hood buckled double, front wheels splayed out and the engine off its mounts, bathed in antifreeze.

The front bumper on the Land Rover was scratched, one fender had a dent in it and there was a broken headlight. That's all.

Brunhilde stood and stared in delighted disbelief, and I said, "Well, I guess that proves the factory's claim that a Land Rover can withstand the full charge of a bull rhinoceros."

"Yes," she answered, "and also the charge of the cow Buick."

While that scene had its memorability, the funniest, (though somewhat dangerous) bit of goofery I ever saw pulled with a Land Rover was on the day the Dude refused to sell us parts.

The Dude was of another common type—short, skinny, big handlebar mustache, tweed cap and bright red vest. They sell used cars. Our Dude was only a bit different in that he owned a big foreign car place, and we'd go up there to get pieces to fix whatever wheezed into our shop. But one day when I rode up with the boss to get some needed bits, we found that the Dude had arbitrarily decided that our place was taking business from his and he wouldn't give us the parts at discount. It was full counter price or nothing.

I'll spare you the shrill 15 minutes that followed. Just suffice it to say that we got back into the Land Rover partless and the boss was so mad he couldn't talk. So after sitting for a moment while he gained enough composure to drive, we headed back out the drive of the Dude's place, and who should be sitting there at the end waiting for traffic to clear, but the Dude himself, encased in a new Alfa. We pulled up behind him.

Four lanes of fast traffic was swarming by and the Dude was watching intently for a hole so he could pull out. We waited. And we waited. And suddenly, the boss reached down and pulled the Land Rover into low range 4wd. I looked over at him. Grinning a fiendish grin, he inched ahead and gently contacted the Dude's rear bumper. The Alfa began to move. The Dude locked the brakes. The Land Rover's engine changed pitch, built to a scream, and with four Pirellis and one Dude shrieking in protest, the Alfa was shoved slowly and majestically out into the middle of the street.

We left him there, the center and cause of an epic traffic jam, and drove away.

Yes, one day I'm going to get one, and it's going to have all the options and attachments I want, too; snow thrower, winch, mower, hydraulic mousetrap, clam digger, twin machine guns, bird call, heavy-duty traffic ram . . . and an air horn that plays "Rule Britannia!"